A REGIONA

THE RAILWAYS (

General Editors: DAVID ST JOHN THOMAS
ALLAN PATMORE

VOLUME I

THE WEST COUNTRY

PUBLISHED IN THIS SERIES

Volume I: *The West Country* (fourth edition)
by David St John Thomas

Volume II: *Southern Enland* (third edition)
by H. P. White

Volume III: *Greater London* (second edition)
by H. P. White

Volume IV: *The North East*
by K. Hoole

Volume V: *The Eastern Counties*
by D. I. Gordon

Volume VI: *Scotland: The Lowlands and the Borders*
by John Thomas

Volume VII: *The West Midlands*
by Rex Christiansen

BY THE SAME AUTHOR

Double Headed: Two Generations of Railway Enthusiasm
(jointly with Gilbert Thomas)

Non-Fiction: A Guide to Writing and Publishing

Summer Saturdays in the West
(jointly with Simon Rocksborough Smith)

(*1*) *Built in the year of the collapse of Gladstone's first government, this 2–4–0 Beattie tank, once the pride of the London & South Western Railway, still had vigour enough to take turns with her sisters on the Wenford Bridge freight line in Cornwall when the first two editions of this book were published.*

A REGIONAL HISTORY OF
THE RAILWAYS OF GREAT BRITAIN

Volume I

THE WEST COUNTRY

by
David St John Thomas

WITH 56 PLATES,
12 ILLUSTRATIONS IN TEXT
INCLUDING REGIONAL MAPS,
AND LARGE FOLDING
MAP

DAVID & CHARLES : NEWTON ABBOT

0 7153 6208 9

First published 1960
Second edition 1963
Third edition 1966
Fourth edition, completely reset 1973

Set in Intertype Baskerville
and printed in Great Britain
by Western Printing Services Ltd
Bristol
for David & Charles (Holdings) Limited
South Devon House Newton Abbot Devon

Contents

Illustrations

PLATES

IN TEXT

Sources of the Photographs

I am indebted to many generous and helpful individuals and firms who supplied me with copyright photographs as follows:

Dr L. N. Owen, plate 2; Mr L. M. Hobdey, plate 3; Mr W. T. Dart, plate 4; Messrs Locomotive & General Railway Photographs, plates 5, 7, 9, 10, 11, 12, 13, 17, 30, 31, 32, 35, 36, 37, 38, 42, 43, 46, 48, 49, 51, 52, 53, 55; Mr R. Russell, plate 6; Mr R. J. Woodfin, plates 8, 54; Mr Albert J. Fellows, plates 14, 24, 47; Mr Roger Sellick, plates 15, 18, 44; Mr J. E. Simpson, plate 16; Messrs R. L. Knight, plate 22; Mr Ivo Peters, plates 23, 50; Messrs Hindley Studios, plates 25, 26, 29; *Illustrated London News*, plates 27, 39; Messrs John Heathcoat & Co., plate 28; Mr Arthur Malls, plate 33; Mr M. E. Ware, plate 34; Mr Nicholas Horne, plate 56. Plates 1, 19, 20, 21 are by the author; plates 40, 41 from the author's collection.

Note on the Maps

The main folding map in three colours is the first ever to show all the railways of the West Country, and special care has been given to the inclusion of minor lines. In order to make the map of permanent value, it was decided not to indicate which stations are closed, particularly as some closed to passenger traffic remain open for goods, and vice versa. The passage of time is ignored, except in one respect: railways whose track was removed by the end of 1958 are shown in broken instead of continuous lines.

The Barnstaple, Par, Exeter and Weymouth areas are enlarged in insets on the main folding map. Maps and diagrams of other complicated areas are included in the text.

The folding map will be found inside the back cover.

Preface

This book, now in its fourth edition, has quite a history and could perhaps be said to be a pioneer work in a number of ways. It was of course the first in the Regional History of the Railways of Great Britain series, of which six other volumes have now been published with as many more in active preparation. Complete coverage should be achieved by the end of the 1970s, but the amount of research and distillation involved is far beyond that needed for most railway books and the project cannot be unduly hurried. It also, may one boast, heralded a new kind of railway history, putting the railways against the general historical background, telling not only how they were built and run, but for what purpose and with what results.

The first volumes in the series were published by Phoenix House, who also published the comparable canal series, The Canals of the British Isles, edited by Charles Hadfield. It was as editors of these two series that he and I first met and later founded David & Charles, which ultimately took over both series. In a sense, then, this was a David & Charles book born in the same year as David & Charles themselves though it had been written two years earlier, publication being delayed by a printing strike.

That there has been a longer gap between the third and fourth editions than between the earlier ones has largely been caused by delaying the massive amount of revision necessary. 1958 proved to be a dangerous if interesting time to be writing railway history that came right down to date, for shortly British Railways were to change out of all recognition. In order to record important closures and other happenings in the second and third editions, it was sometimes necessary to delete other useful material. Now the entire work has been reset and rearranged, and in addition to covering a few more years of history the earlier deletions have been restored, and some other tidying-up done.

Except where otherwise indicated 'the present' means 1973, but I have deliberately left in the two postscripts added for the 1963 and 1966 editions since they seemed to carry a message in their own right, and I have added a similar note for 1973. Since 1958 did indeed prove to be a turning point in the history of the railways of South West England, I have left in many of the original points of comparison labelling them as of 1958 vintage. The bibliography has been entirely recast and inevitably includes quite a few titles published by David & Charles from the railway station at Newton Abbot – offices we moved into when the former running and maintenance headquarters for the Western Region west of Taunton were moved to the new skyscraper block at Plymouth station. The Press were complaining that the building, largely empty, was a white elephant; the railways switched staff from elsewhere to help fill it up, thus giving us the opportunity to work from an address unique in publishing.

CHAPTER I

The General Background

On 29 November 1952, inhabitants of Bridgwater's northern outskirts peered cautiously through their windows. Outside, two men ran to crouch in a doorway, automatically covering their heads as a whistle blew and explosions thudded. An accident at a local armament factory was fresh in memory, and anyway the evening itself was eerie: after a wild, gusty day, snow was falling silently, stealthily subduing the suburb.

The noise was merely a few railwaymen's way of paying their last respects to the Bridgwater branch of the Somerset & Dorset Railway. Detonators had been placed every few yards along the track and the locomotive's whistle was tied down. Residents were taken unawares: for most of them the railway had virtually ceased to exist years ago, and nobody had troubled to spread the news that this would be the last train to set out across Horsey Level and Chilton Moor to Edington Junction.

The only passengers were a handful of railway enthusiasts and one or two reporters. At the Junction we waited for trains in either direction on the Somerset & Dorset's Highbridge line. One arrived empty; the other's sole fare was a red-faced Victorian-looking gentleman soundly asleep. The theoretic connections having been fulfilled, we recrossed the moor and at intermediate Cossington saw the guard extinguish the solitary lamp. The station was closed – in defiance of Parliament. The unrepealed Bridgwater Railway Act of 1882 states that at Cossington, 'The Company shall construct and for ever hereafter maintain . . . a fit and suitable stone or brick building with convenient approaches and proper goods sidings.'

As we alighted at Bridgwater North, quite deserted, snow dripping through the wooden roof, it was hard to believe that when this Act of 1882 was passed church bells rang out in wild rejoicing. Originally the town was served by the Bristol & Exeter Railway, a continuation of the Great Western, whose monopolistic service was not good enough for some of the merchants. They had made three unsuccessful applications to Parliament for powers for a rival route before the triumph of 1882. Friendships were made and broken according to whether families supported the ending or the continuance of the Bristol & Exeter's monopoly. Meetings and petitions were common; so were 'casuistry and cunning'.

The rest of the story is typical. The Bridgwater Railway brought little direct financial reward to its local proprietors but benefited the larger Somerset & Dorset and the still larger London & South Western a great deal. For example, they persuaded Bridgwater people to use the new line to join their route to London (*via* Templecombe) instead of going by the Great Western (with which the Bristol & Exeter had amalgamated). But this competition in itself made the Great Western belatedly improve its service, which eventually far exceeded anything possible by the indirect single-track Edington Junction route. In the automobile age even local patronage of this declined, and the one-time well-supported fourteen daily trains to and from Bridgwater North fell to four carrying a mere handful of passengers. So the Transport Commission decided to risk closure, despite Cossington's 'for ever hereafter' clause. No legal objections were raised.

Economically there was no justification for a second railway at Bridgwater – or there would not have been had the first one provided a good service. As, however, the Great Western would not do this until faced by a competitor, the town does owe the latter a debt. The improved service and reduced goods rates enabled expansion and consolidation at a crucial period.

It was after making that journey in 1952 and reflecting how much more auspiciously the era had begun than it ended that the idea for this book and then for the series occurred – the idea of showing something of what railways have meant in social history, as well as telling the story of their development. The best way to do this is obviously to divide the regions into reasonably self-contained areas, not necessarily corresponding to those served by the individual railways. This approach – it might be called practical or applied, in contrast to academic or pure railway history – will be used throughout the series.

Railways were the very first major mechanical invention directly to affect most of the population, yet much remains to be learnt about their achievements and shortcomings. I hope that this series will help. But, though a final volume summing up the evidence produced from the regions may eventually seem desirable, the authors of the regional volumes are carefully avoiding national history. Apart from other reasons, if greater perspective is required it can at this stage be best supplied by one of the several excellent general economic histories.

And so back to the West Country, my especial concern. Without doubt the arrival of railways was the most far-reaching single peacetime event in the region's history. Our theme inevitably is that of the steady absorption of the region into the main stream of the national life. But even the word region is misleading, for there was no economic, social or other unity about the West at the dawn of the railway age. It was in fact a series of highly individual regions, which had scant intercourse with one another, though most of them were influenced by the sea, depended largely on agriculture and had a remarkably stable population – owing to such factors as the relative freedom from political and other disturbances and the equable climate. Not only have railways integrated the West with the rest of Britain; they have broadened horizons even at the local level, allowing organizations of all kinds to develop on a regional scale. Not that the broadening process must be exaggerated: the West Countryman's conservatism has always exerted strong braking pressure, and today there are probably still more differences – surface and deeper – between North and South Devon, for instance, than between widely separated parts of some other regions. For this reason, once the theme has been set I feel it wise to keep remarks applying to the whole of the West to a minimum and devote nearly all the space to the area-by-area survey.

But first to take a quick glance at the railway revolution in regional perspective. Consider what happened in a single generation, 1830 to 1860.

In 1830, though there were already a few isolated goods railways, particularly in Cornwall, virtually everyone and everything had to travel by sailing ship, on horseback or by horse-drawn vehicle. Several resorts had begun to grow, but for the most part only people compelled to travel did so. The vast majority of workers neither needed nor desired to go afield. The region had largely missed the industrial revolution, and – with one or two notable exceptions – sources of employment had changed little for many years. The owners of most of the small town mills of various kinds felt perfectly secure at the head of their tiny empires. Road towns such as Launceston were expanding steadily. Local and not Greenwich time was still used, while few national newspapers penetrated the far West. Postal arrangements remained crude, although thanks to turnpikes and faster coach services they had improved greatly during the previous decade.

Stage coaches did much to pave the way for the railways, but in the West they never carried heavy traffic, and indeed seating was limited and often passengers had to be refused. The point to be emphasized, of course, is that the coaches had an incredibly short heyday. In 1780 it still took nearly two days to travel from Plymouth to London; and some parts of the West, relying on the traditional packhorses, saw no wheeled vehicles. But by 1830 average speeds of ten miles an hour were fairly common and in 1835 London to Devonport was regularly performed in 21¼ hours. The number as well as the speed of coaches increased – mainly after 1830 and thus within the generation whose beginning and ending we are comparing. The opening of the Liverpool & Manchester Railway in 1830 put speed at a premium everywhere. It speaks volumes, incidentally, that at the end the coaches were timed in half-minutes.

By 1860 the main-line railway system had been completed and its power demonstrated. The places it by-passed, such as Launceston, were already falling back, while a host of new resorts had been created and the holiday trade become firmly established in most of the region as a major source of livelihood (not in Cornwall, exceptional in this and some other respects). Everyone came directly or indirectly under railway influence and a surprisingly large section of the population actually used the trains, if only once a year and mainly for local journeys. Comings and goings between London and the West were at least a dozen times more numerous than in 1830. The increased travelling, the universal

Page 17 From the window in 1951

Plate (2) A 'Grange' class locomotive leaving Cornwall by Brunel's masterpiece, the Royal Albert Bridge at Saltash.

Plate (3) A 'King' and a 'Castle' on an up train halted by signal under the foothills of Dartmoor at Brent.

Page 18 En route to the west

Plate (4) 'City of Truro', 4–4–0 speed victor of 1904, back at work again after its long sojourn in York Museum, takes its enthusiasts' excursion down the 'wrong line' in May 1957 to overtake an engineers' special near Flax Bourton.

Plate (5) A down G.W.R. goods near Frome.

Plate (6) Changing engines on the 'Devon Belle' at Wilton. The 'Devon Belle' theoretically ran non-stop from Waterloo to Exeter, but the lack of water-troughs on the Southern made the unofficial pause for engine changing necessary.

Page 19 A page of broad gauge

Plate (7) Down express at Flax Bourton.

Plate (8) Down 'Cornishman' passing Uphill Junction with Weston-super-Mare loop train in background in May 1892, a few days before the end of the broad gauge.

Plate (9) At Ivybridge, c. 1890.

Page 20 Contrasts in architecture at Exeter

Plate (10) St David's before its last rebuilding.

Plate (11) St Thomas before the lengthening of the up platform.

Plate (12) The dark interior of the old Queen Street.

adoption of Greenwich time, the rising popularity of national newspapers and the greater space given to national events by local newspapers which could now use the telegraph, a most important railway adjunct – all this had begun to blend the West into the rest of the country. The process of standardization and centralization had been set in motion, the closure of many of the small mills whose future had seemed so secure in 1830 being only one indication of the way things were moving. Indeed, most of the trends still in evidence today were firmly established a century ago.

An examination of newspaper contents shows that in many ways 1830 to 1860 was a more definite watershed than the two world wars together. For example, in 1830 there was still relatively little to read about social questions, but even by 1850 the spirit of reform was resolutely on the march, and long columns listed outstanding evils to be remedied. Railways, of course, were by no means the only forces at work: there had been much readjustment and progress from 1815 to 1830, as after any prolonged war. But they were the vehicle of 'creative destruction', the means by which so much of the old order was swiftly swept away.

The West may not have been an industrial area, but its sheer remoteness made railways as essential here as anywhere in Britain. Without mechanical transport it would have been impossible to develop tourism, horticulture and other new sources of livelihood sufficiently to offset the removal of some old industries to the Midlands and North and the decline in mining. But once the railway system had been built, the region – as the rest of the country – could have managed reasonably well without motor vehicles. While the trends the railways established were largely new ones – or as new as any that had been seen in Britain, where evolution and not revolution has always been the order – other forms of transport have merely extended or adapted influences already at work. This is not to minimize the importance of the internal combustion engine, which has, for instance, marked the scenery far more than railways did. Indeed, it could conceivably be claimed that it would have been better had the use of roads remained restricted to horse-drawn vehicles, mainly plying to and from the nearest station. Freedom of movement is highly desirable until the urban sprawl and hideous ribbon development along the coastline is taken into account.

It would be unfair, of course, to compare the types of building stimulated by railways and by motor traffic, for the railway century generally set high store on permanence and ours is satisfied with caravans and concrete blocks. But it is significant that not a single well balanced resort was established off the railway map. The best of the buildings the railways inspired (especially the centres of the resorts, except one or two like Weymouth which had come into their own by 1840) do at least partially offset the damage done to areas immediately round stations, marshalling yards and locomotive depots. Railways are at their grimmest in towns and cities, and likewise the worst parts of the towns are usually seen from them. But outside, constructional sores quickly healed, and the slender tracks, never straight for long and free from buildings on either side, became almost a part of the countryside. To me one of the strongest attractions of West Country railways is the unique range of scenery they open up. Nearly always the view from the train is more rewarding than that from the corresponding road – particularly in Devon, where both ribbon development and high banks restrict the motorist's outlook.

But it is not the purpose of this book to be an advertisement or defence of railways. The story must tell itself, and one of the chief conclusions must be that the railway companies frequently failed to take advantage of their unprecedented potentialities before the arrival of motors. Meanwhile, before starting the area-by-area survey, one or two explanations are necessary, chiefly for the benefit of newcomers to the subject.

From earliest times standard gauge on Britain's railways has been 4 ft 8½ in., but in the West this has generally been known as narrow gauge, as against the broad gauge of 7 ft 0¼ in. The latter was introduced by Isambard Kingdom Brunel, engineer of the Great Western Railway, who conceived all his works on a lavish scale. Originally the GWR ran only from Paddington to Bristol and therefore its first days do not concern this book, but the broad gauge was quickly extended both by the GWR proper and by numerous independent companies. Eventually, of course, it was abolished and standard gauge – what, following West Country tradition, I call the narrow gauge – became almost universal, the only exceptions being some minor lines with gauges ranging from 1 ft 11½ in. to 4 ft 6 in. These genuinely narrow lines are carefully indicated.

Though the first main railways in the West were broad gauge, the earlier independent lines, mostly linking mines and quarries with ports, were 4 ft 8½ in. or less. The 4 ft 8½ in. examples included the Bodmin & Wadebridge Railway in Cornwall, opened in 1834 and acquired (illegally) by the London & South Western Railway in 1847, when that Company's main line from Waterloo was still 200 miles away. The advance purchase of the isolated concern – not for another forty-eight years was it connected with Waterloo – did more than anything else to encourage the eventual spread of the LSWR's narrow system into broad gauge territory, and thus heralded the end of the broad gauge.

The LSWR, second in importance only to the GWR in the West, became part of the Southern Railway in 1923, when several hundred British companies were merged or 'grouped' into the big four – GWR, SR, LMSR and LNER – which lasted until nationalization in 1948. The GWR was the only large railway to retain its identity through the 1923 rearrangements, though the Somerset & Dorset Joint Railway kept much of its old individuality until the 1930s despite its new joint owners.

With the exception of Light Railways (mainly cheaply constructed lines for sparsely populated districts, authorized by Orders made by the Board of Trade) and a few industrial routes for which land was obtained by private treaty, all railways started by an Act of Parliament. 'Authorized in 1876' means, of course, that the Act was passed that year; the exact date of the Act is given only in more important cases. Obviously there has not been space to list all Acts amending the original ones and those needed merely to extend the period Parliament allowed a company for the completion of a line. One of Parliament's ways of controlling railways was not to grant *carte blanche* powers for an indefinite period. Most Acts, incidentally, allowed a company to raise a certain sum in shares plus an additional amount on mortgage at fixed interest rates. Generally, where I speak of authorized capital I refer to the share capital only. One final point about Acts: the largest number were passed in the early 1840s, when unemployment was high and money cheap. The boom in promotion of new lines reached its peak in 1846 and is known as the Railway Mania; it was followed by a sharp reaction in 1847. It has been estimated that 1,000,000 men were employed in building and making materials for new railways at the height of activity, and probably the West had more than its share, at least on the actual construction side. The direct employment had great value in the otherwise bleak 1840s, but research for detailed contemporary evidence has proved disappointing.

An opening date is given for each section, and unless otherwise stated is the day when public passenger traffic began. The official opening often took place a day or more before, and goods traffic usually began later (though occasionally earlier) than passenger business. The openings of all individual stations cannot possibly be included. In later years the GWR was noted for its halts or unstaffed stations. These were mainly provided in two distinct phases, 1905–15 and 1925–30. The second phase was an effort to make good the losses from road competition: ironically, only after a bus service had been started did trains begin stopping at some villages where they would have been far more useful in the pre-motor age. Appendix II shows a detailed list of halts for one chapter area. Closing dates, like opening ones, refer to public passenger traffic. In a few cases goods traffic continues. Junctions are not necessarily stations: a glance at the maps will determine this.

CHAPTER II

From Bristol to Exeter

THE BRISTOL & EXETER RAILWAY

When a company of West Mendip militia fired volleys from flint-lock muskets on the outskirts of a seaside village at the beginning of the nineteenth century, the onlookers – mostly country folk discussing the absurdity of building a big hotel there, let alone celebrating the laying of its foundation stone – could obviously have had no idea that the event would one day be mentioned in a book on railway history. But it is not inappropriate to recall it.

It was nearly forty years later, on 14 June 1841, that the railway reached Weston-super-Mare. By then the place had increased its population from 150 to 2,103, and had become a famed health and pleasure resort, with crescents of handsome houses and plentiful self-assurance. The guide book was firmly established, attracting fresh customers to more than just the first hotel, which – all forty bedrooms – had been opened in 1808.

Flushed with success, Weston felt it could manage without a main line railway: a scheme for carrying the Bristol & Exeter nearer to the sea had been frowned upon, and when visitors first arrived by rail they came down the mile-and-a-half branch in four-wheeled carriages drawn by three horses in tandem, ridden by boys. Except for one daily locomotive, horses remained the rule till 1 April 1851. Then the population was 4,031 and already the town had learned to regret being at the end of a branch.

The railway did not create Weston, but it was able to develop it quickly because the town had first succeeded in bringing itself into existence. Founding a new place was inevitably slow in the nineteenth century, and dependent on individual initiative, such as that of the first hotel builder. Weston's early start in the field, its natural position (even if we do mock at the 'Mare'), its proximity to Bristol, and an improvable railway situation destined it to become Somerset's chief resort.

Between 1850 and 1930, if no longer, the railway dominated the town's economy. Soon excursion trains from Bristol frequently ran along the branch, which had to be doubled in 1866, when a new station was built. Alongside this was a separate excursion platform. The return fare from Bristol was only 1s 6d and as this attracted a different class of person from that more regularly travelling by train, segregation was deemed essential. Weston became used to receiving 'such numbers of excursionists as nevertheless sound surprising to other places'. Outside the station the railway ran an 'excursion hall', where tea was provided at 1½d a cup or could be made with ingredients brought from home. Today many trippers still bring their own refreshments – though they will not find facilities for making tea – and until its closure in 1962 some of those who came by train still arrived at the excursion station (Locking Road, also used by some long-distance expresses for holidaymakers on summer Saturdays). At the end the separate establishment was not, of course, maintained from snobbishness – though the difference between excursion and ordinary fares remained great – but from operational necessity. The rise of Weston is a fascinating story, but it is time to get to grips with the Bristol & Exeter.

Normally the town at the farthest end of a proposed railway took the initiative, but not Exeter. The Company's prospectus was issued on 1 October 1835 by a second group of Bristol merchants encouraged to go West by the first group's Parliamentary success with the London–Bristol Great Western Railway in August. The Company was inaugurated by an Act of 19 May 1836 with £1,500,000 capital and £500,000 additional borrowing powers for a double line from the GWR 'in a certain Field called Temple Mead' at Bristol to Exeter's canal basin, with branches to Bridgwater Docks and Tiverton. Completion of the main line except between Taunton and Cullompton was promised for spring 1838, but not until the meeting of 5 March 1839 was the broad gauge decided upon. The delay was chiefly because speculators – 'irresponsible holders' – had been caught with shares when a sudden national depression in railways occurred. Local *bona fide* investors were not impressed by the promise of a 14 per cent dividend after the opening to Bridgwater, or even by the statement that the first ten miles would 'enable the Company to bring the products of a vast Coalfield' to Bristol. (There was a branch to the now long-disused Nailsea colliery.) By 1841 no fewer than 4,000 of the 15,000 shares had been forfeited.

Eventually the main line to Bridgwater, and the Weston branch (authorized in 1838), were due for opening on 31 May 1841, but the Board of Trade required alterations before public traffic began and, 'one of the engines having been exercised without permission', running into trucks and causing a blockage, even the official opening had to be postponed to the following day. Then Bridgwater, reached non-stop from Bristol by the directors' and shareholders' special, rejoiced for the second time that year: new docks had been opened on 25 March. Public traffic on the railway began on 14 June, when numerous coaches from the South West altered their routes to connect with the trains at Bridgwater, a prosperous, growing town, with a long-established sea trade and an extensive pottery and brick business.

As resources had been concentrated on getting to Bridgwater, the extension beyond was not ready for over a year. With a typical Brunel station of two short sheds with all-over wooden roofs at a respectable distance from the dignified market and administrative town, Taunton officially welcomed the railway on 1 July 1842 and immediately increased its pull over the surrounding country. Both the station sheds were on the down side, the up and down tracks looping with each other as at several other early stations. Beambridge, on the main road west of Wellington, became the railhead and attraction for coaches on 1 May 1843 (the service from Bristol was generous for the era, with eight passenger and two goods trains daily), and a year later to the day Exeter was reached. Then the coaches had only less remunerative cross-country business. The temporary station at Beambridge was closed.

With Brunel as engineer the line was excellently laid out and – a rare occurrence for a trunk route – built without application for additional capital. Between Bristol and Taunton the chief engineering works were the two-mile Ashton cutting and a short tunnel, the shorter Uphill cutting crossed by a graceful road-bridge near the present junction of the western end of the Weston loop, and Somerset Bridge over the tidal Parrett south of Bridgwater. The foundations of the latter slipped and the centering had to be left in position, until in 1843 a timber bridge quietly replaced it. The original was a flatter version of the Maidenhead bridge on the Great Western, a subject of controversy. Maidenhead succeeded, and Brunel's critics never heard of the Somerset failure!

From Uphill cutting across the flat pastures intersected only by the occasional drainage ditch, the line was built dead-straight mile after mile, and even west of Taunton, where curves and gradients were unavoidable, it was designed for fast expresses. To cross the Black Down hills the gradient stiffened to 1 in 163 before Wellington, 1 in 90 to Beambridge and 1 in 81 to Whiteball Tunnel, 1,092 yds long, thickly brick-lined and on a gradient still rising – at 1 in 128. Great was the disruption of life in the surrounding villages when the tunnel (14 shafts were sunk and filled in) and the approaches were being made, over 1,000 navvies being employed. The descent to Exeter is much gentler and presented no trouble.

At first there were intermediate stations between Bristol and Exeter at Nailsea, Clevedon Road, Banwell, Weston Junction, Highbridge, Bridgwater, Taunton, Wellington, Tiverton Road, Cullompton and Hele. Ten more were added during the Bristol & Exeter's independent life, the first being Durston on the opening of the Yeovil branch in 1853. The telegraph was not ready till August 1855 and was worked by the Electric Telegraph Company until the railway took over in 1859.

Hard-pressed for money, the Company had felt unable to buy rolling stock, and the line was leased to the GWR. On average the GWR retained half the gross receipts; with the remainder the Bristol & Exeter could pay only 4–4½ per cent dividends. As the line was built comparatively cheaply and as traffic was heavier than might have been expected, this emphasizes the foolishness of some of the early promises.

In 1845 the GWR felt compelled to propose a more direct route from London to Exeter, and as a polite preliminary it offered to buy the Bristol & Exeter, which would suffer from the competition. The terms (5 per cent on the capital for three years, then 6 per cent, reducible to 5¾ should a narrow line also reach Exeter) were fair and the directors advised acceptance. The advice was firmly rejected at a special meeting on 7 November 1845, when allegations were made that the GWR had been profiteering out of the lease. This expired on 30 April 1849; meanwhile relations with the GWR had gone from bad to worse. Both companies proposed schemes for a route through South East Somerset, and Brunel, finding it impossible to continue as engineer to both, reluctantly withdrew from the Bristol & Exeter.

When the lease ended, the Company, with a mileage of 85½ (including the Clevedon, Weston and Tiverton branches) owned 20 passenger and 8 goods engines, plus a pioneer railcar – a primitive-looking engine and coach combined. The locomotives were designed by Gooch (q.v.), but after 1850 the Company had its own locomotive superintendent, James Pearson (formerly atmospheric superintendent on the South Devon) who designed various classes, including perhaps the most ponderous tank locomotives ever built, 9 ft single double-bogies, the great driving-wheels being flangeless. One of this class attained 81·8 m.p.h. descending Wellington bank. In 1859 the Bristol & Exeter's locomotive works at Bristol turned out its first complete locomotive; in 1870 it produced its first tender passenger locomotives, 2–4–0s with 6 ft 7½ in. driving-wheels. At the end of its career the Company owned 95 broad and 28 narrow gauge engines, of which 20 and 7 respectively were built at Bristol. Until 1860 the engines were painted dark green picked out with black and then became black all over. Except for the 'railcar' (*Fairfield*, sold in 1856) they were un-named, and class numbers were not kept in sequence. (There were also two 3 ft gauge locomotives for the Westleigh quarry tramway, a mile long from Burlescombe station, opened in 1875; its gauge was standardized in 1898.) To complete the brief rolling-stock digression before returning to the general story, most of the coaches and trucks were built at Eastover works, Bridgwater, where a considerable staff was employed. The coaches resembled the GWR's, but the first and second-class compartments had distinctive oval windows – though to begin with the seconds had no windows at all. Convertible coaches – narrow bodies on broad frames – were built by the Bristol & Exeter before the GWR, but we shall see that the Company did little else to prepare itself for the coming gauge change. Livery was dark crimson lake, usually seeming dirty black. In 1864 the first-class carriages cost £650, third-class £400 and coal waggons £120.

After it had settled to its independent existence, by the standards of the day the Bristol & Exeter was a well-conducted and generally popular railway, even if its passenger fares were about the highest in the country and the minimum of service was provided for third-class travellers. By far the largest of the independent companies in the West, with a capital of nearly £6,000,000, it eventually worked 213½ route miles (actually owning 138½ miles), linking places as far apart as Clevedon and Yeovil, Wells and Barnstaple. Like the GWR it was a pioneer in the smooth running of expresses, regularly attaining speeds unknown on the narrow gauge. Though remaining closely associated with the GWR, it gradually developed a character of its own – a good character but spoilt by the vexatious gauge question. The Bristol & Exeter consistently spent too much of its energy in the attempt to defend its territory from the invading narrow gauge. In this it was unsuccessful; eventually it had to lay the rails for narrow gauge trains on its own main line and two of the broad gauge branches it fostered (Glastonbury and Crediton) were completely lost to the 'narrow' cause.

In 1867 the Somerset & Dorset Railway wanted to reach Bridgwater and was encouraged by the Corporation. The only way the Bristol & Exeter could defeat the scheme was by first adding a narrow gauge rail to its main line from the junction with the Somerset & Dorset at Highbridge through Bridgwater to Durston and then along the branch to Yeovil. £125,000 was spent, and in November 1867 a daily narrow goods train from Yeovil began arriving at Bridgwater. The narrow passenger service was delayed by complications at Yeovil (q.v.), but from November 1868 there were several passenger and 'mixed' workings from Yeovil to Bridgwater or Highbridge.

Had the Bristol & Exeter used the third rail efficiently, probably no more would have been heard of the need for an independent narrow route to Bridgwater. As it was, however, every effort was made to discourage merchants consigning goods *via* Highbridge and the Somerset & Dorset. Though empty broad waggons were always to hand at Bridgwater docks, narrow ones required for through traffic had to be ordered up to a week ahead. No matter what traffic was waiting to go to the Somerset & Dorset, only one narrow goods a day would be run up to Highbridge –

and this remained so even when, with the opening of the Somerset & Dorset Bath extension in July 1874, there was for the first time a through narrow gauge route from the Midlands to Bridgwater (*via* Gloucester, Mangotsfield, Bath, Evercreech Junction and Highbridge). The result was that in 1875 some of the town's merchants again pressed for a second railway, and though the Bridgwater Railway Bill was defeated that year such enthusiasm had been roused that another attempt was made and succeeded, the Bridgwater Railway – the first mentioned in this book – eventually being opened in 1890. Not all the merchants shared in the 1875 excitement: some were wise enough to see that despite its gauge complex the Bristol & Exeter had done much for Bridgwater. It was claimed that £200,000 was spent on works there.

A Corporation horse tramway – the 'Communication Works' – from the station to a wharf on the Parrett, was taken over in 1863 (it had been opened in 1845) and converted into a mixed gauge locomotive line in November 1867. In March 1871 it was extended by an £8,000 bridge across the river to the docks, bought with the bankrupt Bridgwater & Taunton Canal for £64,000. The port became fifth in Britain for the import of coal, the total trade more than trebling in the decade ending in 1874. That year no fewer than 159 boats were owned in the port, and one coal merchant alone distributed inland by rail 40,000 tons of the 60,000 tons he imported. Some of this would have been from Dunball wharf, lower down the Parrett. The wharf and a tramway connecting it with the main line were built by the merchant in 1844 and later taken over by the railway, which extended the wharf and reconstructed the tramway as a mixed gauge locomotive line in 1869. The coal merchant, and a brick manufacturer dispatching an annual average of 10,000 tons by rail, were generous in their praise of the Bristol & Exeter before the Select Committee examining the Bridgwater Railway Bill. The total railway tonnage handled at the town was stated to have increased from 117,000 tons (yielding £27,000) in 1863 to 200,000 tons (yielding £40,000) in 1874.

This must have gratified James Cresswell Wall, whose name is prominent in the Bristol & Exeter's history. The energy and enthusiasm he put into his work while canvassing Bristol for trade for certain Midland canals was brought to the notice of the directors, who appointed him traffic superintendent. Wall apparently found his new position too confining, for in 1855 he left the staff to become, to quote MacDermot's *History of the Great Western Railway*, the Company's 'Agent for the collecting and delivery of goods and parcels and the general management of goods, including the provision of clerks and porters and responsibility for claims, on commission under a contract.' In 1870 he returned to the staff as general manager, but retained his agency on the commission basis. Superior officers had no rigid salary scales in those days.

Nor in those days did management brook interference. When a Bridgwater porter sued the Company for wrongfully dismissing him (alleging he was expected to be on duty 14 hours a day, save 30 minutes for lunch), he had himself represented by a union. The railway's representative was reported in the Press as saying that the Company 'don't intend, never did intend, to be interfered with or dictated to by a self-appointed authority, who might wish to – for the purpose of being clothed with some brief authority – and choose to interfere in some matter in which they were not concerned.' The porter lost his case after it had been proved that he was absent when a slip coach arrived. Slip coaches were inaugurated at Bridgwater in 1869, and later added at Yatton and Tiverton Junction.

After Wall became manager, use of the narrow rail between Highbridge and Yeovil decreased, not increased. Five of the eight locomotives specially built to work on it were temporarily converted to broad gauge. The folly of this escapism was suddenly brought home with the opening of the Somerset & Dorset Bath extension, already mentioned. Yeovil, Exeter, Plymouth and other places as well as Bridgwater now had a route without break of gauge from the Midlands, and immediately the Bristol & Exeter lost much traffic. Its only course was also to provide a continuous narrow route to the West by laying a third rail on the parts of the main line still exclusively broad, between Bristol and Highbridge, and Durston and Exeter. The first narrow train from Bristol to Taunton and back ran on 30 May 1875, regular traffic starting in time for the Royal Agricultural Show at Taunton in July; the Weston branch was also made mixed. Narrow gauge goods first reached Exeter in March 1876, but passenger trains were all broad till July the following year. There was, of course, a shortage of narrow rolling stock.

So far the Bristol & Exeter had been more prosperous than the GWR, but what was now in any case a serious position became desperate when in November 1875 the Somerset & Dorset was leased by the Midland Railway and the LSWR jointly, the two big companies being able (and willing) to pursue a more active policy in developing through traffic than would the Somerset & Dorset had it remained independent. On 23 September, Wall wrote to J. Grierson, general manager of the GWR, setting in motion a series of negotiations resulting in the end of the Bristol & Exeter. The Company was leased to the GWR from 1 January 1876, and amalgamated with it from 1 August. Shareholders received a guaranteed 6 per cent increasing to 6½ per cent (terms similar to those offered years before), to pay which the GWR temporarily had to reduce its own dividends. (See Appendix I for detailed statistics for one year – 1861.)

MAIN LINE FROM 1876

Though nobody seems to have worried unduly about such details as lighting signal lamps at night – until a crash (at Cheddar) the rule book omitted the customary instruction that invisible signals should be regarded as at danger – the Bristol & Exeter was a safe railway. Interlocking was introduced generally on it long before other West Country lines. No passenger was hurt in the Company's one serious accident, at Creech, near Taunton, in 1852, when a locomotive suddenly left the road at high speed. But no sooner had the GWR leased it than two disturbing smashes occurred, though again without loss of life to passengers. On 24 May 1876 the leading coach of the down *Flying Dutchman* left the rails near Silverton. Speed was about 60 m.p.h. – as it was at Long Ashton on 27 July the same year, when the engine of the up *Dutchman* jumped the rails, hit the down side embankment and toppled over on the down track. Fortunately the coupling broke and the train ran on along the up line. The track, as well as being weakened by the addition of the narrow rail, had been disgracefully neglected at the end of the Bristol & Exeter's life. So with public confidence somewhat shaken, the GWR had to impose a 40 m.p.h. speed limit and spend a small fortune on relaying. The Board of Trade also insisted on the lengthening of most station platforms: few could accommodate even ordinary trains. The railway's less sentimental officers must have wished that the opportunity was taken to abolish the broad gauge there and then.

Broad trains were now exceptional east of Bristol, but farther west the change was effected slowly. The first frequent main line service of narrow gaugers ran to Weston; one by one the former Bristol & Exeter's branches were narrowed. Then, on 1 March 1884, the Weston branch and the Junction and town stations were closed and replaced by the present loop and through station. The loop was exclusively narrow gauge, and as the resort (with a population of 13,000) was too important to be omitted by stopping trains between Bristol and Taunton, these now had to be narrow. After this, indeed, the broad gauge was confined to through trains to Devon and Cornwall – apart from coal traffic from Dunball and Bridgwater to the Taunton–Exeter district. The daily Dunball coal train to Exeter remained broad until the end of Brunel's gauge in 1892. An increasing proportion of seaborne coal now came into Bridgwater itself; but the whole coal trade in the West Country was gradually changed after the opening of the Severn Tunnel in 1886. Bridgwater docks, improved in 1907, more or less held their own, but other harbours down the Bristol Channel suffered as merchants were persuaded by the GWR's often uneconomically cheap terms to switch to the rail. The process has continued down to the present. Until about 1948, for instance, diminutive boats crossed the Bristol Channel to bring the small place of Porlock Weir, on the Exmoor coastline, its coal; but now this travels by rail to Minehead and then by road. Since the war in particular, South Devon and Cornwall have also been receiving more of their coal by rail. Coal is therefore an important freight on the Bristol–Exeter main line. Even between Bristol and Taunton the railway generally is much busier now than at any time last century, although all London to West Country traffic used it then. Since 1906, of course, most of the traffic between London and Devon and Cornwall has taken the Castle Cary route, to be dealt with in Chapter VIII, but even between Bristol and Taunton traffic continued to increase and necessitated many new works.

Bristol is outside the scope of this book but it may be noticed that under the Development (Loans, Guarantees and Grants) Act of 1930 the GWR and LMSR were able to borrow money for numerous improvements, including electric signalling at an extended Temple Meads and the quadrupling of the track to the divergence of the Portishead branch beyond Parson Street. This permitted a better train service to the south. One and a half miles to the west and to the east of Yatton were quadrupled in 1921 and 1925 respectively: it had been intended to quadruple through the station, but the cost of land was deemed too much. Originally the present Weston General station had two through platforms, with one excursion terminal platform some distance away. A bay has been added to General, while the excursion Locking Road was rebuilt as a spacious four-platform terminus.

At first the Castle Cary line joined the Bristol route at Cogload by an ordinary junction. At the beginning of the 1930s, however, the four miles from the junction through Taunton to Norton Fitzwarren were quadrupled and a bridge was provided to carry the down Bristol track across the two from Castle Cary. The Bristol & Exeter had replaced the original two-shed Taunton station by a larger single unit with an all-over roof in August 1868; in 1895 the GWR had trebled the platform lengths, added bay platforms, and opened a goods-avoiding line to the south. The station was completely rebuilt during the quadrupling, being completed in February 1932. Also during the early 1930s the track through the stations at Wellington, Sampford Peverell, Tiverton Junction, Cullompton and Stoke Canon was quadrupled, only the loop lines serving platforms.

Many of these improvements have since been abandoned. Reductions in train services (100 main-line winter passenger trains daily in 1958 had come down to 70 in 1963 and in 1973 there are only 63) made Yatton a less serious bottleneck; the quadrupled track was gradually removed until with the extension of the area controlled by a new power signal box at Bristol only a solitary goods loop was left. As part of the new signalling, Bristol Temple Meads itself was reduced in size, while sidings installed near Flax Bourton in the 1950s had meanwhile been removed. Weston-super-Mare Locking Road was closed in 1965, and in 1972, also as part of the signalling scheme (Bristol controls as far as Cogload Junction), the Weston loop was singled except actually at the station. The quadrupled track west of Taunton now stops at Silkmill Crossing, short of Norton Fitzwarren, and most of the loops added to stations between Taunton and Exeter in the 1930s had been pulled up by the end of the 1960s. Some of these simplifications were made possible by the increased speed of vacuum-brake fitted goods trains, and also by their reduced length since banking engines are no longer used and loads are thus confined to what a single diesel can haul up Wellington Bank. For a period before the new signalling came into use mineral trains from South Wales supplied foundation material for the Bristol–Bridgwater section of the M5 at a number of temporary sidings.

Weston-super-Mare's population grew from 34,039 in 1921 to 43,923 in 1961 and 50,794 in 1971; residential development continues apace, in parts extending inland from the main line. Even between the wars the railway was a primary factor in helping the town grow as a resort. Through weekend trains ran from many Midland, Northern and Welsh towns as well as from London. Day and half-day excursions from Paddington were introduced in 1905 and 1925 respectively. Summer weekends still bring the occasional special excursion and several thousand people still travel by regular services on fine Sunday afternoons, but nothing like the pre-war bank-holiday total of 38,000 will ever be seen again. The bombing of Bristol helped give Weston a new role as dormitory town and this the railways encouraged after the war. In 1958, eight trains, five of them starting from Weston itself, arrived in Bristol before 9.30 a.m. This is now down to a sorry four, though two more – including an early Taunton–Bristol–Paddington express – race up the main line, missing the town. Weston has indeed not been served generously or consistently in recent times, and one suspects that the increasing number of trains not calling is due partly to the singling of the loop line, and that in turn was done only because big savings had to be demonstrated to get Government approval for the £4,000,000 Bristol power signalling. Weston now goes all out to attract light industry, but until it is totally rebuilt the town centre cries aloud that here is a railway creation.

Bridgwater is another place larger than many people think and is still growing. Its 1971 population was 26,598. It was not well served in broad-gauge days when the dock branch was often clogged with traffic, and this century it became something of a depressed town. A third of its workers were unemployed in 1933, and even immediately after the war the outlook was bleak. But then energetic local efforts succeeded in attracting a wide variety of industry. The biggest factory, alongside the main line and with its own private railway system, is the British Cellophane Company's. A short narrow-gauge tramway from a clay pit to the brick works passed under the main line west of the station until 1958. Once there was considerable railway employment in the town, but little remained after the closure of the former Bristol & Exeter's carriage and waggon works, and in January 1967 even the rail connection with the docks was severed.

Taunton station was once served by local trains from seven different routes, and all day long loading and unloading of mail, parcels and even livestock took place in the numerous bay platforms. There was also a full-scale daily commuter cycle. Though the track is still quadrupled through the station and a non-stop train can overtake a stopping one, only a single through platform is now used in each direction, plus one bay for starting up trains. True, the town has a faster and more frequent service to London, the far West, Bristol, the Midlands, and the North than ever before, but it has long forgotten that its role as Somerset's capital was consolidated by the local rail traffic. It also owed much of its earlier growth – the population has doubled since about 1880 (1971: 37,373) – directly to railway work, which has of course declined sharply, though the first freight and coal concentration depots in the South West were opened here in 1964.

Between Bristol and Taunton, principal stations have always been served by some express as well as less important trains. Between Taunton and Exeter the pattern was very different: until the closure of all intermediate stations except Tiverton Junction on 5 October 1964, virtually all trains stopped everywhere or not at all. Only since then and since the closure of the Tiverton branch has Tiverton Junction enjoyed a balanced if limited express service with trains to and from Paddington and the Midlands. The pressure of through traffic used to prevent Wellington, Tiverton Junction and Cullompton enjoying the kind of service they would otherwise have justified, though a summer-only semi-fast to the South Devon coast did brisk business especially at Wellington during its main mill's holiday fortnight. West of Norton Fitzwarren (more recently Silkmill Crossing) there is but one pair of rails to carry what has previously used two – the separate Bristol and Castle Cary routes to Cogload, and then the quadrupled track. Many were the holidaymakers who reported they had a clear run to around Taunton and then seemed to stop at every signal.

We end the section, as we began it, with an account of accidents. On the night of 11 November 1890 a slow, down narrow train assisted by a broad pilot engine shunted on to the up track at Norton Fitzwarren to enable a fast down goods to overtake. This was a common procedure, there then being few refuge sidings or goods loops. Nor were there any mechanical or electrical safeguards to remind the signalman of the presence of the shunted train or to stop him signalling another train into it. Ten of the 50 passengers on an up Plymouth ocean special were killed when, at high speed after descending Wellington bank, it ran into the slow train. That was the last major accident on the GWR for half a century – fifty years minus seven days, when again the scene was Norton Fitzwarren. On the night of 4 November 1940 the down sleeping-car express was running on the relief line from Taunton. The driver, however, read the main line signals, which were in favour of a newspaper train. Only when this began to overtake him did he realize his mistake – too late to prevent his *King George IV* being derailed where the quadrupled track ends and six of the coaches being strewn across all four tracks, killing 27 of the 900 passengers. The newspaper train cleared the spot in time, sustaining only superficial damage from hurtling ballast.

BRANCH AND MINOR LINES

CLEVEDON owes much of its character to its railway's early arrival, on 28 July 1847. Broad gauge until 1879, the branch's traffic has remained fairly constant, though the modern tripper usually by-passes Clevedon, which lacks good beaches and as a holiday town has seen its heyday. For many years there was a shuttle service of about 30 daily trains to Yatton; many were well patronized. But the service declined before closure from 3 October 1966.

Clevedon was the headquarters of the WESTON, CLEVEDON & PORTISHEAD Railway, opened to it from Weston on 1 December 1897 and extended to Portishead on 1 September 1907. A steam tramway had been proposed, but a full-scale railway of standard gauge was built; in 1899 it became a Light Railway. In 1904 the original Company was taken over by another: eventually both went into liquidation, a receiver was appointed, and the railway was operated by the late Colonel H. F. Stephens, an almost legendary figure who acquired a number of Britain's independent but bankrupt lines and through sheer enthusiasm kept them running well into the motor age. For thirty years the Light Railway's sole creditor was the Excess Insurance Company, which withdrew its petition at the beginning of the war. The railway should have been returned to its owner, but as neither company could be traced and the receiver was no longer entitled to an interest, trains ceased from 19 May 1940. The Government refused intervention because 'the closing would not gravely impair the district's transport facilities', the truth of which statement cannot be denied, even though on a few occasions each summer nearly all the heterogeneous rolling stock was pressed into action: normally trains called, or would on request, at 17 intermediate places, mainly level crossings, along the 14 single-track miles, and railcars were used extensively. Several thousand pounds await the owners of the railway if they are ever discovered, but to settle its debt the Insurance Company claimed the track (not the goodwill) and sold it to the GWR, which hoped to use it for storing loaded coal waggons. But the GWR had no engine light enough, and none of the Light Railway's was available, so the line was dismantled.

For its size WELLS, with three stations within a third of a mile radius each built by a different railway, has one of the most complicated histories. The Somerset Central first proposed placing it on its main line from Highbridge toward Frome, but then chose to develop southward from Glastonbury to meet the Dorset Central, which it joined to form the Somerset & Dorset. Wells was to be by-passed but, owing to pressure from shareholders representing the city, a branch was built, this being opened on 15 March

1859 – actually before the main line was extended southward. The gauge was broad, then mixed, and narrow after 1868 (see Chapter X). The East Somerset Railway was opened from Witham to Shepton Mallet on 9 November 1858 and extended to Wells on 1 March 1862. Last but most important in the long run was the Cheddar Valley & Yatton Railway, opened to Cheddar on 3 August 1869 and to Wells on 5 April 1870. Its Act had been obtained by the Somerset & Dorset, but next year powers were transferred to the Bristol & Exeter, which in return dropped a Weston-super-Mare to Wells (*via* Bleadon) scheme. This would have robbed the Somerset & Dorset's main line from Highbridge of much traffic. The Cheddar Valley and East Somerset lines were broad, with the narrow Somerset & Dorset separating them at Wells. The Somerset & Dorset agreed to allow broad gauge trains across its property and a rental was fixed. But the Board of Trade refused to comply on safety grounds: a sharp exchange followed when the Somerset & Dorset pointed out that the rental had to be paid none the less. In 1874 and 1875, respectively, the East Somerset and Cheddar Valley lines were standardized and exchange of goods between the three railways began. But only on 1 January 1878 did the GWR (which had now absorbed both the East Somerset and Cheddar Valley) run passenger trains over the nine chains of Somerset & Dorset separating its two sections. The same day the former East Somerset station was relegated to goods, and Tucker Street became the only station used by GWR trains, though they passed through the passenger platform at the Somerset & Dorset's Priory Road. It took fifty-six years for prejudice to die: Priory Road was used jointly from 1 October 1934 till its closure for passengers with the Somerset & Dorset branch on 29 October 1951. So, for passengers, only Tucker Street (then just Wells) remained until the closure of the Yatton–Witham line to passengers from 9 September 1963. Traffic on this line had always been heavier on the Yatton than the Witham side, the Cheddar Valley Railway quickly producing gross receipts of £20 per mile per week. Excursion traffic from Bristol to Cheddar and Wookey (opened 1871) was once brisk, though by 1950 had been nearly all lost to the road, which had also gained much of the strawberry traffic – the fruit is grown on 250 holdings. Once almost every station had its milk churns to load: before the war the one Sunday train used to convey eight vans of milk for London. A remnant of the district's once famous paper industry provided some freight traffic until 1964 when the branch was entirely closed except for quarry traffic on a short stump between Witham and Cranmore. But this limestone business has grown dramatically and new installations include a 1½-mile branch to Merehead and extensive sidings at Witham.

Farmers in the area west of Congresbury (first station from Yatton on the Wells branch) found marketing their dairy produce difficult, and encouraged by the Light Railway Act and of course not suspecting that the motor age was at hand, persuaded some local landowners to help them promote a six-and-a-half-mile railway to Blagdon. The scheme was supported by Bristol Corporation, about to construct its large Yeo Reservoir, and the GWR agreed to work the line. The WRINGTON VALE Light Railway was opened on 4 December 1901 and apart from carrying equipment for the reservoir in its early days it led a quiet rural existence for many years. Bus competition caused economies in the early 1920s, and the passenger service was withdrawn on 14 September 1931: goods traffic continued to decline, and on 1 November 1950 the line beyond Wrington was closed completely and the remainder on 10 June 1963.

CHARD fully expected to be placed on a trunk route connecting the Bristol and English Channels, but suffered from being by-passed by the main east-to-west routes. Schemes to convert the canal, opened from Taunton in 1842, into a railway failed, and only on 8 May 1863 did the first train arrive – on the branch from Chard Road, which became Chard Junction. The LSWR had already absorbed the local Company. But people were more interested in the line to Taunton. Powers for this were obtained by another local Company in 1861 and transferred to the Bristol & Exeter in 1863. Running parallel with the canal and also serving the small town of Ilminster, Chard's second branch was opened to a joint station on 11 September 1866. The LSWR reached Chard Joint (later Central) by using powers for a loop from its Chard Town to the canal basin, though the canal – already nearly derelict – was bought by the Bristol & Exeter and immediately closed. At Chard Joint the two railways, of different gauges, kept aloof: each had its own signal-box and staff. On 19 July 1891 the broad gauge was at last converted, the GWR deliberately postponing it through fear that the LSWR would press for running powers into Taunton. But rivalry was replaced by the need for economy. Chard Town, until then still used by many LSWR trains, closed on 1 January 1917, when the GWR undertook to work Chard Joint to Chard Junction. Both lines were closed to passengers in the fuel crisis between 3 January and 7 May 1951, and finally from 10 September 1962.

Today Brendon Hill, inland from Watchet and 1,200 ft above the Bristol Channel, is as quiet a spot as any in England, but in the 1850s it was alive with iron-mining activity. In 1855 the WEST SOMERSET MINERAL Railway was incorporated with £65,000 to connect the workings on the Hill and at Gupworthy beyond with Watchet harbour. Construction began in 1856, and next April limited ore traffic began from Comberow, at the foot of three-quarters of a mile of 1 in 4 cable-operated incline, which carried the line on to the Hill. The incline was opened temporarily in 1858, and with balancing equipment added, public goods traffic officially began on 28 September 1859. On 24 June the Ebbw Vale Company of South Wales leased the railway for 55¼ years at £5,595; its successor, the Ebbw Vale Steel, Iron & Coal Company, adopted the lease, obliging itself to keep the line in good order till 1919. Passenger traffic to Comberow began on 4 September 1865. Until 1882 mineral traffic was heavy, most months reaching 3,000 tons, transferred to ships at Watchet harbour, where after 1861 the quays accommodated 500-ton vessels. Suddenly Welsh smelters found it cheaper to import Spanish ores, the mines collapsed, and the 400 miners and their families left; even the railway's general traffic was reduced. The passenger service was cut to one daily train and on 7 November 1898 the line closed, though the rental continued to be paid. When another mining boom began, the Somerset Mineral Syndicate sub-leased the line and on 4 July 1907 re-opened it from Watchet (where £1,500 was spent on a new jetty) to Brendon Hill, starting point for a two-mile 2 ft gauge mine tramway. The Gupworthy extension remained closed. In 1910 the mines were abandoned, the jetty auctioned for £70, and a £2,000 kiln built for refining ore at Washford sold for £5. Next year the Watchet–Washford section was used for an Australian firm's demonstra-

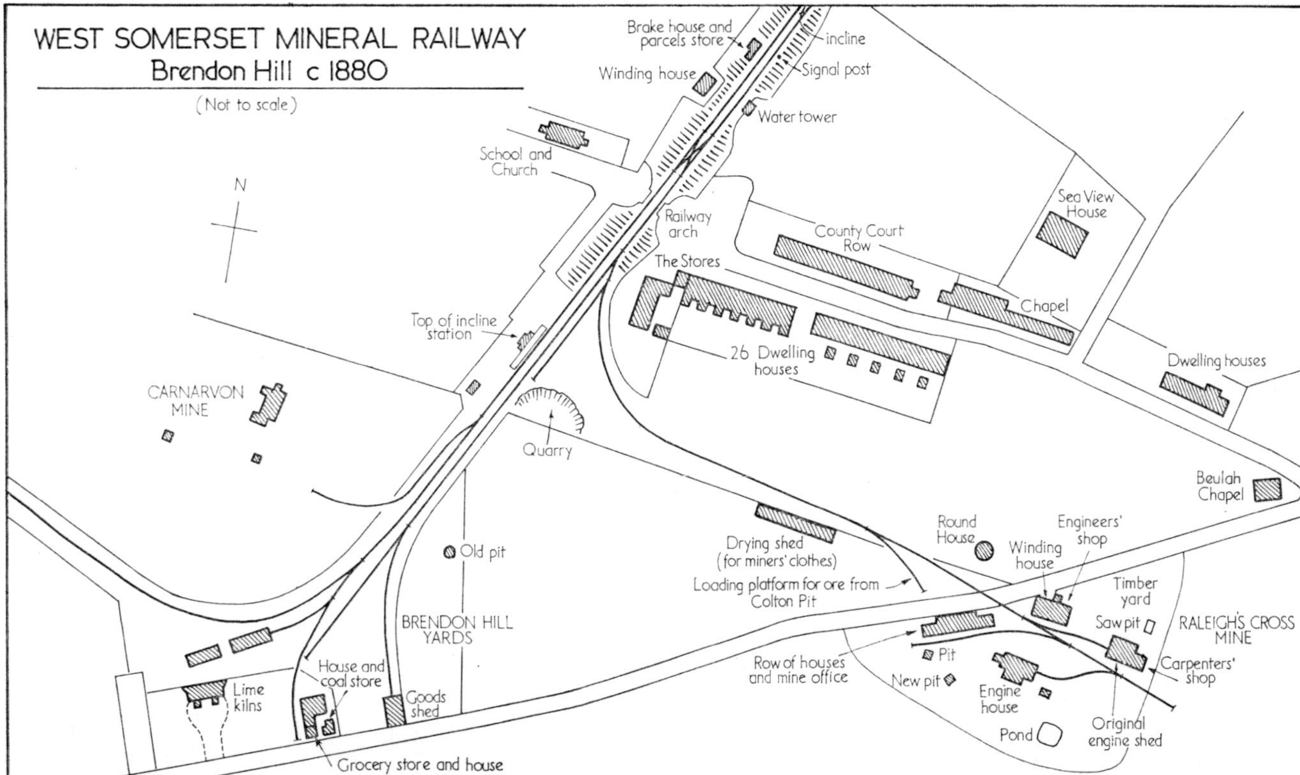

1. Grocery store and house. 2. House and coal stores. 3. Drying shed for miners' clothes. 4. Row of houses including mine office. 5. Loading platform for ore from Colton Pit. 6. Engine house. 7. Winding house. 8. Engineers' shop. 9. Original engine shed. 10. Carpenter's shop.

tions of automatic train control; otherwise only cursory maintenance was done till the rails were commandeered in 1917. Most of the Railway Company directors then wanted an abandonment Order, but one objected and established a rival Board: no railway, two Boards and a steady income from the unfortunate Ebbw Vale Company till 1919, when the lease expired.

The MINEHEAD branch was built by two companies. Watchet harbour was the objective of the West Somerset Railway, authorized in 1857, started in 1859, but because of difficulty in raising the £160,000 not opened (from Norton Fitzwarren) till 31 March 1862. The extension, the Minehead Railway, was authorized in 1865, dissolved in 1870, revived in 1871 and opened on 16 July 1874: there had been serious labour trouble, navvies once being kept out of the office of the clerk of works only at the point of a loaded revolver. Both were worked by the Bristol & Exeter; the Minehead became part of the GWR in 1897, but the West Somerset retained independence till 1922. The gauge was converted in October 1882. Until well into this century traffic was light: there was no permanent connection with the mineral line at Watchet, and though that town's people nobly rebuilt their harbour after gales of 1900 and 1903 (the Urban Council was formed to meet the situation) there was little rail distribution of seaborne cargoes. But gradually Minehead's population grew (less than 2,000 when the railway arrived, it was over 6,000 in 1931) and the surrounding villages became tourist-minded. In 1933 the crossing loops were increased from four to six, in 1934 automatic token exchangers were provided and Dunster–Minehead doubled, and in 1936 Norton Fitzwarren to Bishop's Lydeard was doubled. Freight ceased under the Taunton concentration scheme, and a number of ingenious economies were made, such as singling the Bishop's Lydeard section and extending the single-line branch through Norton Fitzwarren to a new junction with the main line at Silkmill Crossing. The Dunster–Minehead double track was left in position but operated as two separate single lines on a one-engine-in-steam basis enabling Minehead signal box to be closed. At the end only two signal boxes remained, while traffic, helped by the Butlin holiday camp opened at Minehead in 1962, was brisk; that the Minister of Transport authorized closure came as a shock to railway officials as to local inhabitants. Closure was from 4 January 1971. At the time of writing the track remains in good condition and reopening under private enterprise is under discussion.

The Devon & Somerset was an unfortunate railway: when absorbed into the GWR in 1901, its arrears of debenture interest amounted to £500,000, and twice (1868 and 1889) shareholders suffered through rearrangement schemes. The Act for the long line from Norton Fitzwarren to BARNSTAPLE was passed on 29 July 1864. The 1866 crisis intervened, and it was 8 June 1871 before the market town of Wiveliscombe was reached, and not till 1 November 1873 that the first train ran to Barnstaple. The line was cheaply built with mean stations and only three crossing loops in 43 miles of single track. During the construction of the first part, provisions ran out and seventy navvies marched to Wiveliscombe armed with bludgeons and demanding bread and beer. At first the train service was scandalous; mails were not carried because of a disagreement over terms, and coal was prohibited from travelling up from Barnstaple. The Bristol & Exeter, which leased the railway, expected to deliver coal imported at Bridgwater to South Molton, but the town's merchants boycotted it and for some time continued collecting their stocks by road. Improvements came with more crossing stations, the conversion to standard gauge in May 1881, and the opening of the spur to the LSWR at Barnstaple Junction on 1 June 1887. Through expresses to Ilfracombe were instituted, and carried increasingly heavy traffic. Station loops were lengthened, automatic token exchangers provided, and Norton Fitzwarren–Milverton doubled in the mid-thirties, when through Ilfracombe traffic reached its peak, with up to half a dozen expresses on top of the ordinary local service on the busiest Saturdays, a restaurant car down to South Molton and even a Sunday train. Myxomatosis robbed South Molton of its daily rabbit special, and in 1963 the once-heavy cattle traffic also ceased. Next through freight to Barnstaple was diverted *via* Exeter; then the whole branch closed to freight; then even summer Saturday trains *via* Taunton to Ilfracombe were sent *via* Exeter, leaving only single-car diesel units till complete closure from 3 October 1966. Though some economies were made at the end, the branch was expensive to maintain, with three tunnels and bridges over the Tone, Exe and Bray (the first and last with wrought-iron girders on stone piers), 12 signal boxes, and frequent 1 in 60 gradients restricting loads to 9 passenger coaches or 28 goods waggons. But it served a wide area – in spring artificial manures once arrived in hundreds of tons – and was a colourful local institution that is missed.

The EXE VALLEY branch, between Morebath Junction, on the Devon & Somerset, and Stoke Canon, was built in two parts. The Tiverton & North Devon Railway was authorized in 1875 and opened – its trains running through to Dulverton – on 1 August 1884. The Exe Valley Railway, authorized in 1874, handed its powers to the Bristol & Exeter in 1875. It was built by the GWR and not opened till 1 May 1885. Both lines were standard gauge. The Exe Valley's passenger service was meagre in the early days, but vastly improved at the end of the 1920s, though trains were slowed by having to serve five new halts. (Until 1894 Stoke Canon station was east of the junction with the branch.) Although sizeable crowds arrived on special occasions, most trains were lightly loaded and the whole Exe Valley branch was closed completely on 7 October 1963 except that grain traffic continued between Stoke Canon and Thorverton for another year.

The Bristol & Exeter's TIVERTON branch was opened early in railway history: on 12 June 1848. Originally it was of course broad gauge, but it was standardized in June 1884, in readiness for the Tiverton & North Devon. A new station at Tiverton was opened at the same time. Tiverton Junction–Tiverton always had a shuttle service, connecting with both main line and Exe Valley trains, and carried most of Tiverton's once substantial freight traffic. Passenger trains were withdrawn on 5 October 1964, and freight ceased on 5 June 1967.

Finally the CULM VALLEY, from Tiverton Junction to Hemyock, which obtained powers in 1873 as a Light Railway, under the little-used provision of the 1868 Regulation of Railways Act. Landowners and farmers eagerly subscribed to the project, and early and cheap completion followed by large profits were promised, the Bristol & Exeter undertaking to work it. The total expenditure was £50,000, not under £30,000 as promised; the opening was not at the end of 1874 but on 29 May 1876; and the traffic was less than half that expected. Disillusioned, the landowners met for the last time on 2 April 1880 and sold the erstwhile pride of the valley to the GWR for £33,000. Taking a broad view the expenditure was justified: a textile firm was encouraged to open by the railway, and milk production is on a prodigious scale compared with what it was last century. Passenger trains were withdrawn on 9 September 1963, but milk traffic from the creamery at the terminus continues, this being the only branch dealt with in this chapter whose rails remain intact.

CHAPTER III

From Salisbury to Exeter

AN EPIC PARLIAMENTARY BATTLE

> On one of the most unpropitious days of the spring of 1856 the first sod was turned of an undertaking whose previous history had been a career of unexampled trial, and whose future progress was to be still more remarkable for success. In keeping with the bitter elements of turbulent hostility which the Salisbury and Yeovil Railway undertaking had had to encounter, but at this time had seemed to have subdued, were the deluge of rain, and the bitter blasts of wind, which flung it into the faces of the people who were going to the town of Gillingham to be present at the ceremony of 'turning the first sod'. It seemed as though, human adversaries having done their worst to impede, Nature had now taken up the work of baffling and obstruction . . . but the promoters of the undertaking had had too much of the buffeting of adversity to be depressed now . . . in truth, theirs had been a stormy infancy . . . for when local obstructions had been overcome natural difficulties fell across the line. The Wars on the Indian Frontier, the Potato Famine, the French Revolution of '48, the war between Russia and Turkey, all these were impedimenta to the railway. 'We proved our case always,' said Mr Henry Ker-Seymer, the Member for Dorset, at the opening banquet. 'We proved the case, but never got the line. We were unfortunate. Whenever we got a chance there was sure to be a Commercial crisis. The Central line – Potato failure! The Central line! – Money at 7 and 8 per cent! The Central line – Deficient harvest. The Central line! – Wheat at £20 a load.'

Thus begins Louis H. Ruegg's *The History of a Railway*, published in 1878 and recently reprinted under the title of *The Salisbury & Yeovil Railway*. Its opening merits quotation partly because the book is scarce and the language has its own appeal, and partly because it is well to remember the delaying influence of national crises such as Ruegg mentions. In the space available here, the account of the Company's birth-pangs must confine itself to the rival railway schemes.

Even in the 1830s there had been plans for a 'direct' (i.e. not *via* Bristol) railway from London to Exeter; but we may begin in 1845 when three schemes came to the fore. There were then already two embryo narrow gauge routes to the West, a devious one to Salisbury and one nominally direct but in fact winding to Dorchester. The latter, promoted by a Wimborne solicitor, A. L. Castleman, won the appellation 'Castleman's Corkscrew'. Independent companies deposited Bills for the extension of each of these lines to Exeter. The Board of Trade had indicated it favoured a direct Exeter route and, as already seen, the GWR felt compelled to propose a broad gauge scheme.

'Castleman's Corkscrew', the Southampton & Dorchester Railway, once had GWR support, but this was withdrawn as part of a conciliatory agreement with the LSWR in January 1845. In return, the LSWR undertook not to go west of Salisbury or Dorchester. However, no sooner did the GWR announce its direct Exeter line than the LSWR retaliated by saying – without detailed explanation – that the agreement had been broken. Therefore, as these were the days of the Railway Mania, it too decided on a drive (from Salisbury or Dorchester or both) to Devon's capital. It was in 1845, incidentally, that the LSWR illegally acquired the small Bodmin & Wadebridge Railway in Cornwall. The proposed extension from Salisbury to Exeter was called the Central scheme, and already there was another narrow gauge Central scheme to continue the route through the heart of Devon and Cornwall to Falmouth. The Central cause won wide support, though Yeovil quickly protested when it realized it was to be served merely by a branch. The town's property would decline by 40 per cent, it was complained.

But the LSWR was not committed to the Central Salisbury–Exeter route, and the alternative Dorchester, or Coastal, route gained favour. In 1846 both schemes were thrown out by Parliament – as also the GWR's, opposed by the Bristol & Exeter as well as the LSWR. In 1847 the two narrow schemes and the direct Exeter GWR came up again – plus the Bristol & Exeter's South

East Somerset line, proposed in opposition to the GWR's. The GWR met defeat; while the Bristol & Exeter's scheme, the Central (the LSWR's Salisbury & Yeovil) and the Coastal (the nominally independent Exeter, Yeovil & Dorchester) – the two were to be linked at Yeovil – were all passed by the Commons that year and by the Lords in 1848. Both the narrow gauge Bills received Royal Assent on 22 July 1848.

The 1847 fight before the Parliamentary Committee lasted fifty-three days and is reputed to have cost between £300,000 and £400,000. Witnesses for the rival causes were rushed to London on the slightest pretext and treated royally while waiting to give their evidence. Engineers of little experience were paid large sums to swear that one line was perfect and its competitor badly planned. Conveyancing and other routine legal business in East Devon and much of Dorset came almost to a standstill, solicitors finding the railway work far more rewarding.

But it was of no avail, for by the time the Acts had been passed the reaction to the Mania had set in, and international happenings made it increasingly difficult to raise money. Meanwhile the LSWR dividends were falling. Nothing happened till 1851, when an independent company for each of the narrow schemes was formed. Even an independent concern would, however, need the moral support of the LSWR, and the LSWR was a company divided. Some shareholders preferred the Central, some the Coastal, some wanted neither. At first the majority were for the Coastal – now known as the Dorchester & Exeter Coast Extension Railway – and the Central collapsed; then the Coastal fell down on Standing Orders and the Central – the Salisbury & Yeovil – issued its prospectus claiming LSWR support. The LSWR directors were nine to four in favour of granting this and expected confirmation at the meeting of 26 October 1852. But the voting was 12,610 to 12,389 against making any arrangement with the Salisbury & Yeovil.

Hoping to confuse the position further, the GWR now brought forward a Devon & Dorset Railway from Maiden Newton to Exeter. As a 'direct' line it was useless, Exeter being ten miles farther from London than *via* Bristol, and there can have been no intention but to delay the narrow thrust into the West. Action being needed to keep the GWR at bay, the LWSR shareholders firmly authorized their Board to guarantee a double-track trunk coastal route. Upon the new chairman, Francis Scott, conveying this pledge to the Select Committee, the Devon & Dorset application was rejected. The LSWR announced: 'A solemn engagement has been entered into by the Company, and it is alike politic and just to fulfil it.' But at the next meeting 13,914 votes were recorded against proceeding with the Coastal project, three more than for honouring the undertaking.

This was a vote of no confidence. Scott promptly resigned and five directors left the Board. *The Times* congratulated the shareholders on having a will of their own, but then had to pay damages in a libel action brought for remarks reflecting on Scott's character. Meanwhile, Scott's predecessor, W. J. Chaplin, a less aggressive personality, was back in the LSWR chair. Soon another meeting decided to support the independent Salisbury & Yeovil, which had retained faith in itself through these vicissitudes. Its Act received Royal Assent on 7 August 1854.

But no more happened than after the first Act had been obtained for a Salisbury–Exeter railway in 1848, and while population and prosperity continued to fall alarmingly in East Devon it was felt that no railway would ever mature. Fortunately for the district, the Government – wanting better transport facilities between Plymouth and Southern England's other naval and military establishments – had different views. In 1855 the LSWR went to Parliament with an application for extra time to finish its proposed Andover–Basingstoke line, needed to shorten the distance between London and Salisbury. To its amazement, the occasion was used roundly to tell it to honour its pledge to build a narrow route to Exeter. In his report to the Board of Trade on the Railway Bills of 1855, Captain D. Galton wrote that having achieved the aim of defeating the Devon & Somerset Bill,

> the London & South Western Railway Company repudiated their pledge. Consequently when the Company brought forward a Bill for other objects in 1855, the House of Commons referred the Consideration of the Bill to a Committee of nearly the same Members to whom the Devon and Dorset Railway Bill had been referred in 1853. This Committee inserted clauses into their Bill by which the Company were bound, under the penalty of stoppage of their dividends. to introduce into and

use their best endeavours to pass through Parliament a Bill for a narrow gauge line to Exeter.

A 'narrow gauge line to Exeter' could mean either the Coastal or the Central, and early in 1856 it was announced 'that an intimate connection between the South-Western and the Salisbury & Yeovil Companies is now established'. By the summer considerable progress had been made on the Salisbury & Yeovil works. Then a question was asked in Parliament about the liability of the LSWR not only to go to Exeter but to do so as had first been pledged – by a double-track Coastal route. Work all but ceased on the Salisbury & Yeovil, being built as a single-track route though with land enough for a double line. But the Coastal scheme was dropped finally, on 21 July 1856, the LSWR received its Act for the Yeovil–Exeter line, and the Salisbury & Yeovil was assured of its place as 'a bit of the backbone of the great direct Western system'. Because of its interests west of Exeter, the LSWR would almost certainly have reached the city one day, but it would obviously have preferred to take its own time – and perhaps leave the pioneering to an extension of the Salisbury & Yeovil or to another independent company.

DEVELOPMENT OF THE MAIN LINE

The Salisbury & Yeovil was opened from Salisbury to Gillingham on 2 May 1859, from Gillingham to Sherborne on 7 May 1860, and from Sherborne to Yeovil (meeting the Bristol & Exeter at Hendford) on 1 June 1860. Salisbury is just outside this book's area and is included in Volume II, though we may note that the city's original Milford station was closed and Fisherton opened the same day as the first part of the Salisbury & Yeovil. Details of Templecombe and Yeovil will be found in Chapters X and VIII respectively. Yeovil remained the terminus for only a few weeks, the LSWR from Yeovil Junction to Exeter being opened on 19 July 1860.

The railway runs from east to west, parallel with the sea and against the grain of the valleys. The only valley followed for any distance (12 miles) is that of the Axe, and taking advantage of this involves a detour: when it leaves the river four miles from the sea the line has to turn sharply inland again. For most of the route a switchback was inevitable. Even where there are no marked valleys the West Dorset–East Devon country is steeply undulating.

The see-saw gradients begin at Semley, 18 miles from Salisbury. After a dip to Gillingham comes a rise to Buckhorn Weston Tunnel (where water in Greensand caused trouble, delaying the opening to Sherborne), then a dip ending midway to Templecombe, a rise through Templecombe – the straight line presents a spectacular appearance from the platform – to Milborne Port and a dip to Sherborne. Here an Act had to be obtained for a diversion – or rather to cancel a diversion away from the Castle, the new owner not being so hostile to trains as his predecessor. Climbing begins again at Yeovil Junction and after two short dips is resumed in earnest with a long section of 1 in 80 through Crewkerne to the 133¼-mile post, whence follows the steady decline beside the Axe to half-way between Axminster and Seaton Junction. A 1 in 100 rise to the junction becomes 1 in 80 for 4½ miles beyond (speed drops, giving time to enjoy the hill and wood scenery) and climbing continues even through Honiton Tunnel, 1,353 yd. To Broadclyst the descent is then almost continuous, much of it at 1 in 100; three short rising sections are taken in stride by the accelerating expresses.

Despite the difficulty of the terrain and though at first having only a single line, the railway was excellently laid out. Except for the doubling, it has needed little attention and in the steam age speed restrictions were virtually non-existent. Yet it was cheap to build, especially the Salisbury & Yeovil part. Including the original eight stations and the telegraph, but making no allowance for the Templecombe Junction branch, the Salisbury & Yeovil cost but £12,513 a mile. Doubling cost another £3,884, making a total of only £16,402 a mile for a first-class trunk route. The land was cheap, partly because the line traversed few places of any size. The protests ignored, Yeovil was relegated to a branch; Shaftesbury, on a hill, was left without a railway. No doubt this discouraged some local traffic, though from the start the line was busy. Doubling works were begun as early as 1861. The whole route from Salisbury to Exeter was doubled by 1 July 1870; most of the original bridges had been built for two sets of rails.

Ruegg says that when the first sod of the Salisbury & Yeovil was cut at Gillingham, on 3 April 1856, the Company had only £4 2s 4d in the bank. The financial tide turned quickly – particularly after the Company asserted itself by taking the LSWR to arbitration over non-payment of rebate for through traffic originating between Salisbury and Yeovil. This rebate had been fixed as part of the initial agreement leasing the railway to the LSWR, but the latter held that the clause was no longer effective as under a new agreement the line was leased for a straightforward 50 per cent of gross receipts. The Salisbury & Yeovil won, the LSWR having to pay over £3,000 in rebate. Following this there seems to have been difficulty in making a fresh agreement. After several changes the LSWR undertook to work the railway for the exceptionally low figure of 42½ per cent of gross receipts. The 57½ per cent enabled the Salisbury & Yeovil to pay its first dividend, in the first half of 1861. The figure for the whole of that year was £4 7s 6d per cent. This slowly increased until £6 5s or £6 10s became the rule, in 1866–70. Then came the phenomenal rise. In 1871 £6 15s was paid; in 1872 £7 15s, in 1873 £8 5s, in 1874 £8 15s, in 1875 £9 15s, in 1876 £11 5s and in 1877 £12 10s. Nothing equalled this in the annals of railways.

Unfortunately the Salisbury & Yeovil's success led to much false optimism. Promoters of other lines felt that if only they could overcome initial difficulties similar prizes would follow. But the line's circumstances were peculiar. Though traffic was heavy, little additional capital expenditure was needed, no important branches being undertaken. Up and down the country many railways failed to pay good dividends through overstretching their resources. The contrast between the compact Salisbury & Yeovil and the sprawling Somerset & Dorset intersecting it at Templecombe could not be clearer. Moreover, while the Somerset & Dorset was financially exhausted after completing its costly Bath extension in 1874, the Salisbury & Yeovil, without bearing any of the expense, shared in the through traffic resulting from the new line. (The Bath extension completed a continuous narrow gauge route from the Midlands to the South West.) The interchange of both freight and passengers at Templecombe was considerable.

Under the terms of the final lease, the LSWR was obliged to commute the 57½ per cent of the gross receipts paid to the Salisbury and Yeovil into a fixed yearly payment twenty years after the opening to Yeovil. Rather than wait till 1880, however, in 1872 it offered to exchange £100 of the smaller Company's stock for £150 of its own 5 per cent preference stock. The Salisbury & Yeovil directors urged acceptance, but led by Ruegg himself the shareholders rejected the terms. In 1877 the LSWR offered £250 of its ordinary stock for £100 of Salisbury & Yeovil; 'judicious coyness' resulted in the final offer of £260 (either ordinary or 5 per cent preference) per £100. This gave Salisbury & Yeovil shareholders the choice of ordinary stock then saleable at £347 1s per £100 nominal capital, or a perpetual preference dividend of 13 per cent. The terms were accepted and the smaller Company vested in the larger from January 1878.

For the rest there is little to tell except that the route continued to be improved as it became busier. Templecombe, once a frustrating bottleneck, was given a new station with extensive marshalling facilities in the early thirties. Yeovil Junction was rebuilt in 1907 and like Seaton Junction, rebuilt in 1928, had quadrupled track, the platforms being served by loops.

On top of generally-increased traffic in the 1914–18 war, the wayside station of Dinton, eight miles from Salisbury, suddenly sported great activity. In 1915 an enormous military camp was opened at Fovant, about two miles away. At first transport between station and camp was provided by traction engines, panting and wallowing in mud; but soon the Fovant Military Railway was opened, and for the rest of the war and a year or so afterwards – when the camp was a demobilization centre – prodigious traffic was handled. Shunting operations at Dinton often seriously delayed regular trains, and timekeeping suffered throughout the LSWR's West Country system.

The passenger emphasis has always been on the long-distance aspect, discussed in Chapter XI. Exeter's influence stretches surprisingly little to the east, and when in 1906 a rail motor service was introduced between the city and Honiton, it met with little success and was soon curtailed. Many more workers have always been carried into the city by the Exmouth branch than by the main line. In 1958, however, no fewer than six stations were the starting and terminal points for local trains, most of them to and from Exeter, though this perhaps demonstrated the lack of a clear traffic pattern more than the volume of business. A very small Exeter commuter business remains, and the former Sidmouth Junction has been reopened under its original name of Feniton to accommodate workers.

Yeovil (population in 1961: 24,552), with its shops and aircraft, glovemaking and other industry, is capital of a larger area, but much local traffic was lost to the road as early as the 1920s, though Yeovil Junction remains the busiest station between Salisbury and Exeter. From the traffic point of view it was of course a pity that Yeovil and most other important towns were served by branches, and that the junctions of all these branches except for that of the Lyme Regis line at Axminster were way out in the country at points with little potential of their own. All the branches dealt with at the end of this chapter except the Exeter–Exmouth line have been closed, and so have all the junction stations except again Axminster and Feniton. In all cases the branch trains had their own bay platforms at the junctions.

At one time electrification through Salisbury all the way to Exeter seemed probable, and it could of course still take place between Basingstoke and Salisbury. Today's traffic certainly will not justify capital works beyond Salisbury, and the question must be whether the route has a long-term future at all. Unfortunately the combination of singling the track and using an unreliable class of diesel locomotives led to frequent major delays (once pasengers in a train broken down beside the A30 near Honiton became so frustrated that they walked to the road and hitch-hiked) almost literally driving some business away. The route was singled, with long sections between crossing loops, except between Gillingham and Yeovil Junction; that double section was left to ease the passing of up and down trains at speed, but poor timekeeping frequently defeats the aim. The locomotives were the Western's first diesel hydraulic class, the 'Warships', unreliable enough when new and ever more troublesome as they aged. They have been replaced, but by less powerful if better-behaved machines and at the cost of lengthening many journey times. Here is a route where you really could travel a lot quicker in steam days.

But it was perhaps the withdrawal of the railway from the centre of Yeovil that did as much as anything to reduce traffic.

Various alternatives had been discussed to reduce the number of stations at Yeovil, including the retention of only Yeovil Town as a terminus at which all through trains by both routes would have had to reverse. Though this would have slowed through traffic, Yeovil was accounting for an increasing proportion of business by both routes and, apart from keeping the station in the town centre, the arrangement would have made transfers possible between the two lines. To go from the South West to Weymouth now means buying a ticket via Westbury.

Goods traffic was never on the same scale as on the Bristol–Exeter line: for one thing little coal was carried, though some coal and Burton beer for the South West was routed via the Somerset & Dorset; even into the 1950s transfers of freight at Templecombe were heavy. Otherwise agricultural produce, including cider from Wimple, were the heaviest loads. All through freight traffic ceased shortly after the Western took control as part of the 1962 regional boundary adjustments, and now only milk traffic is handled, that from the creameries at Chard Junction and Seaton Junction being worked to London *via* Exeter. In 1939 the Southern was carrying 9,300,000 gallons of milk from Semley, Sherborne, Chard Junction and Seaton Junction; by 1957 this had doubled to over 18,000,000 reflecting the productivity of the rich countryside through which the route passes.

As it winds and climbs the main line seems part of that countryside – a land which, because of the railway's late arrival, has been spoilt remarkably little. By 1860 the greatest drift from the land had ended and much of the damage to the economy was irreparable. Sherborne regained some prosperity, but not its former importance; Gillingham's trade improved, but only at the expense of Shaftesbury's; Axminster had lost its carpet industry long before the first train arrived (since 1937 carpets have again been made there, in a factory beside the station); and Honiton was reawakened only by the motor-car. Before the railway era the local roads were vastly superior to those of most of the West, and Honiton was a thriving crossroads town. It was thus especially hard that the whole area should be left virtually off the transport map between the opening of the Bristol & Exeter to Exeter, in 1844, and 1860.

BRANCH LINES: THE EAST DEVON COAST

Moreover, 1860 brought only the opening of the main line. Lyme Regis, just in Dorset, and all the East Devon resorts, were left to be served by branches, which followed slowly. The resorts lost visitors to places served by trains, their fisheries declined, and no new industry came to replace the once famous but rapidly dwindling East Devon lace trade. An 1887 inquiry into the trade's plight was told that during the preceding generation the number of lace workers at Beer had fallen from 400 to between 60 and 70, and at Exmouth none of the one-time 20 lace schools had survived. The distress of the poorer population was appalling.

All the resorts clearly display the branch-line stamp. Had it been geographically and economically possible to link them with a main line along the coast (the so-called Coastal route proposed from Dorchester turned inland before reaching Devon), they would be larger today and have lost individuality. The small places, such as Beer and Branscombe, between the resorts would probably have degenerated into sections of ribbon development. One may be glad history took the course it did, but enjoyment of the bold, rounded cliffs is marred by the thought that they stand unspoilt only because the district's economy was once brought to so low an ebb, causing widespread suffering. Even had the inland main line been opened by 1850 and the branches added within the next five years, it would have made a big difference. When assessing the influence of the railway's late arrival on the resorts, however, it is only fair to add that shingle beaches, unpopular among family holidaymakers and day trippers, are the rule in East Devon, except at Exmouth. At Sidmouth in particular the lack of sand has had profound importance.

LYME REGIS was already changing from a port to a resort when in 1845 the Railway Mania produced three schemes including it; but nothing was done about building a line even after a Lyme Regis Railway cut its first sod in 1874. In 1898 Chard and Axminster people were still bitterly arguing about the relative merits of Chard Road and Axminster as starting-point for a branch: next year a local company settled the point by obtaining the Axminster & Lyme Regis Light Railway Order. Concrete for the ten-arch Cannington viaduct (one arch quickly had to be supported by a jack arch, which spoilt the appearance) began arriving at the Cobb, Lyme's man-made harbour, in 1900. The Cobb's remaining shipping business rapidly declined after the railway's opening, on 24 August 1903; but it was the picturesque harbour and cottages overlooking it – one of Jane Austen fame – which attracted visitors in increasing numbers. At Axminster the branch crossed the main line by a flyover and had a bay platform on the up side. A goods connection on the down side was removed in 1915. The only intermediate station, Combpyne (where there was a crossing loop until 1930), once handled an extensive traffic in sightseers going to the cliff landslip, of 1839 vintage, between Lyme and Seaton. During six months of 1908, crowds also used the branch on their way to see a cliff fire between Lyme and Charmouth. In the previous year the local Company had been absorbed by the LSWR, soon obliged to relay the branch, which carried 60,000 passengers annually. Road competition was felt seriously – Lyme station was perched awkwardly above the town – and only on summer Saturdays was there an economic quantity of work for the three survivors of Adams's 4-4-2 radial tanks, withdrawn in 1961, which for many years provided all motive power. The line was closed on 29 November 1965; goods traffic had already been withdrawn.

SEATON is a modern town, the places on either side formerly being bigger. Axmouth's once-flourishing harbour was reviving from a bad spell when the independently-owned branch from Seaton Junction was built down the other bank of the river. Trade disappeared and ten years later the river-mouth was bridged, cutting off the port. Beer had elaborate harbour plans, but with the continued absence of safe anchorage the substantial fishing-trade declined – and what is left is still declining. With the collapse of lace-making as well, the Beer that visitors like for its quaintness today was during much of last century one of England's most poverty-stricken large villages. Harbour plans at Seaton itself also came to nothing, so the railway, opened on 16 March 1868, has always been chiefly concerned with tourism, though even this developed slowly. By 1900 the population was only 1,300; those accustomed to taking holidays preferred better-equipped Lyme or Sidmouth, and the awkwardness of the rail journey from any centre but London delayed growth when holidays with pay became common. It was only when many working people began to own cars, and through motor coach services started running from the Midlands and North, that Seaton came into its own. But on summer Saturdays up to 500 long-distance passengers (nearly all from London) arrived at the station, rebuilt by the Southern Railway in 1937, and there was some excursion traffic. The line closed on 7 March 1966; again freight traffic had already been withdrawn. A miniature electric tramway has been built on the trackbed from Seaton to Colyford.

Seaton's station was on the sea-front and beside a large holiday camp; SIDMOUTH is a mile inland and even the bird flying above it would spot no caravans. Sidmouth deliberately discouraged the railway from coming near enough to please trippers and, aiming to remain select even in the second half of the twentieth century, it now actively discourages camping. But, as at Seaton, there were harbour plans. The first locomotive in the town was narrow gauge, bought for a temporary line in connection with an 1836 dock venture: it is said to have been too big to go through a tunnel. £12,000 had been spent when the scheme was abandoned, and another £7,000 was lost on a similar scheme linked with the Sidmouth Railway, authorized in 1862. The Company collapsed after some of the line's earthworks had been built; it was revived in 1871 and the line opened on 6 July 1874, the harbour now being forgotten. Sidmouth had for eighty years been attracting a select number of visitors, specially for winter residence (a theatre was established as early as 1805), and the coming of the railway made less difference than at any other resort in the West. Traffic down the Otter valley from Sidmouth Junction to Tipton St John's (which as junction for Budleigh Salterton and Exmouth once had over 50 trains a day) and then over the steeply graded section to the resort was never heavy, but remained steady, the town's growth from 4,033 population in 1901 to 11,139 in 1961 – there is a high percentage of retired people – offsetting the loss of some classes of business to the road until the closure on 6 March 1967 (two months later for goods). The local Company incidentally found it worth retaining independence until 1923.

EXMOUTH is another old-established resort. Like several in Devon it developed quickly during the Napoleonic wars, when the Continent was closed. The population reached about 5,000 by the end of the 1840s: then the South Devon Railway trains could be seen on the other side of the river carrying the trade to Dawlish, Teignmouth and Torquay. Exmouth declined: the ferry to Starcross ran in some kind of connection with train services, but the crossing in rowing-boats did not allure those who could enter one of the other resorts by rail. In 1855 the Exeter & Exmouth (an earlier company of that name had come to nothing) was authorized as a broad gauge line to leave the South Devon Railway at Exminster, but the LSWR wooed the Company and persuaded it to build a narrow line and share Queen Street. The line was opened on 1 May 1861, and was of course worked by the LSWR, which absorbed the Company in 1866. After the opening new building was resumed in Exmouth, and in 1864 the docks (now the only active commercial ones on the East Devon coast though no longer rail-connected) were opened. Coal used to be imported and distributed by rail, and once pilchard specials used to run to London; but like the other branches in this chapter, its emphasis was always on passenger traffic, probably heavier here than on all the other lines together. Even in the early 1880s commuting business into Exeter was considerable; by the first war it had been found necessary to double the branch from Exmouth Junction to Topsham, as far as which rail cars, calling at a number of halts serving Exeter's suburbs, augmented the train service. In 1963 a 'short' train to Topsham still supplemented the 31 down Monday to Friday Exmouth trains. These included a non-stop commuter service introduced in 1960. The branch was officially removed from the Beeching list of closures, but traffic has declined since 1957 when tickets collected at Exmouth totalled 486,457. Exeter Central's figure was little more: 529,457. There are still 20 Monday to Friday trains, but only one platform of Exmouth's four-platform terminus built between the wars is now needed and all signals have been removed beyond Topsham – where the spirit of the LSWR perhaps lives on better than anywhere else. The short Topsham Quay branch was removed after being little used for a long time in 1958. Freight traffic had ceased on the Exmouth branch itself at the end of the previous year.

In 1892 steamboats were introduced on the Starcross ferry. Six years later the ferry was bought by the GWR and for many years

'Starcross for Exmouth' did mean something. Regular passenger traffic is now scarce, but each summer thousands of excursionists buy joint ferry-rail tickets to South Devon resorts.

BUDLEIGH SALTERTON, a small resort and residential town which has much in common with Sidmouth, welcomed its locally-sponsored Budleigh Salterton Railway, which continued following the Otter from the junction with the Sidmouth branch at Tipton St John's, on 15 May 1897. The LSWR, which took over the local Company in 1912, built the extension to Exmouth, opened on 1 June 1903. Much of the through London–Exmouth traffic was diverted to the new line, and several through trains round the circular Exeter–Exmouth–Budleigh–Sidmouth Junction–Exeter route were inaugurated. Use of all these East Devon branches was encouraged by the introduction of runabout tickets just before the 1914–18 war. They were also used by moderate numbers of London day trippers after about 1900 and until 1939, and even in the 1950s there were three summer Saturday through services to London and one to Cleethorpes via the Somerset & Dorset. Freight ceased in 1964 and passengers on 6 March 1967.

CHAPTER IV

Exeter and South Devon

EXETER

When the railway arrived in 1844, Exeter had just undergone the most thorough revolution in her history. Empty warehouses told the fate of the famous clothing industry which had once employed four-fifths of the city's workers. The population had jumped by over 10,000, partly from an influx of village weavers, unemployed as the woollen business moved north. No alternative work existed in country areas and though the city offered little enough, hundreds had opened new shops or set up as petty craftsmen. A model sewerage system had been built and cholera was not to be seen again; only twelve years before, the Exeter of dark alleys and foul smells, still largely confined within the old walls, had been devastated by the scourge.

These immediate pre-railway years had brought other changes, not the least important being the development of the stage coach business. Not only had the amount of travelling greatly increased since the beginning of the 1820s, but speed had crept up until the Exeter–London run was performed in the astonishing time of 16½ hours. But on its opening day – 1 May 1844 – the railway proved its superiority: the directors' special, hauled by the locomotive *Orion* and driven by Daniel Gooch, left Paddington at 7.31 a.m., reached Exeter at 12.45 p.m., left again at 5.19 and reached Paddington at 10.3. (See Appendix III.) No wonder Thomas Latimer of the *Western Times* was enthusiastic, devoting a four-page supplement to the opening and claiming 'that increased intercourse with the more populous districts of England cannot but prove highly advantageous to the fair and lovely spinsters of Devon'. St David's station, well beyond built-up

Exeter, consisted of 'two elegant buildings in the drabesque style of architecture'.

Latimer was spokesman for progressive Exeter. He had long appealed unsuccessfully to the wealthy to encourage the railway by investing in it. The Church, the town's most powerful voice, was indifferent; vested interest in the roads and their coaches was strong. After the opening, of course, the coaches were left only west-bound traffic. The Plymouth road became livelier than ever, the journey to Devonport being performed in four hours 'for months together'.

It was due to Latimer that an appreciable number of working people first used the railway: a club was formed to organize Great Exhibition trips. Otherwise Exonians had little immediate need to travel on the Bristol & Exeter (though the whole city came under its influence) and were more interested in the South Devon Railway, opened two years later, which provided easy access to the sea. The city's third railway, the Exeter & Crediton, opened in 1851, will be described in the next chapter. Then, in 1860, the LSWR's route from Waterloo was at last completed. From 1861 its Queen Street terminus was also used by Exmouth trains, and next year it ceased to be a terminus. The 50 chain connection at 1 in 37 (partially through a tunnel) down to St David's was authorized in 1860 and opened 1 February 1862.

The LSWR's down trains travelled on the Bristol & Exeter's up track through St David's and the mile beyond to Cowley Bridge Junction, where the Crediton line began. Thereafter, up trains to Paddington and down ones from Waterloo travelled in the same direction through St David's. The old station being inadequate for the extra traffic, a new one was started in 1862 and opened in July 1864. With four platforms, an impressive ornamental frontage and a partly-glazed roof 132 ft wide, 363 ft long and 34 ft above rail level, it was heralded as a wonder of the age.

Though the South Devon passed within half a mile of Exeter's ancient Ship Canal, a branch to the City Basin was not opened until 17 June 1867. Narrow rails were laid on it and on the main line (down track only) and LSWR trucks reached the waterside from 1871. Trains of both gauges' trucks were run.

As the West's most important railway crossroads, Exeter quickly became the centre for the surrounding country, but for a long time was too dependent on that country's agriculture to be healthy. The plight of Devon's farming toward the end of the century was serious; bad outbreaks of cattle disease in the mid-1860s not only caused immediate loss but, worse, delayed the necessary transition from crops to cattle. As little new industry arrived, it was only when the city developed as a shopping, distributive, administrative and educational centre for seaside as well as agricultural Devon that the economy became reasonably well-balanced. By the turn of the century the population was approaching 50,000 and the amount of coming and going by train was above average even for a county town of that size.

In 1906 the LSWR opened three suburban halts east of Queen Street: Lion's Holt (now St James Park Halt), Mount Pleasant (closed 1928) and Whipton Bridge (closed 1923). The same year powers were sought for an independent line from Queen Street to Cowley Bridge Junction: there was keen competition for the Plymouth ocean traffic, and the LSWR was maddened by GWR insistence that all 'foreign' trains should observe a stop at St David's. But the competition soon ceased, the idea of a separate line was dropped, and then the present St David's was built (1911–14), only the old frontage being retained. There are five platforms, plus a through down track and a bay for Exe Valley trains; Western goods trains by-pass the station. St David's is more than a mere railway station. On its platforms have been enacted many historic scenes; its atmosphere is strongly individual, and from that point of view one may be glad that a 1935 scheme to give the Southern a separate high-level station did not mature. Work started – the Exe was channelled away from the station's rear to give more room – but was halted by the war and has not been resumed. Until 1964 Southern trains continued to use the centre island and tank engines to whistle shrilly when announcing their readiness to 'bank' trains up to Exeter Central. St David's remains busy, with an hourly service from Paddington from 7 a.m. to 7 p.m., plus a night train that detaches a sleeping car, and an almost hourly service from Birmingham and Bristol. The two-hourly trains to Waterloo start here, as do the locals to North Devon; for several years the down bay platform, previously the home of the Exe Valley auto-car trains, served the single diesel car for Okehampton.

In 1958 I wrote that St Thomas, Exeter's 'overhead' station, with its dank, sloping corridors, 'wears a neglected air – save for the imposing South Devon Railway frontage of 1846', which was given a face-lift when the offices were let as shops in 1959 – and added that 'the overall roof is singularly incongruous now that the platforms have been much lengthened'. A battered look would be a more appropriate description today; not merely has the overall roof gone, but parts of the main building (seen more clearly during an urban renewal scheme in the Exe Bridge area) have been knocked down with no consideration for appearance. Only five Monday to Friday trains now stop in each direction.

The original Queen Street, dark as a crypt, was altered several times. Bay platforms were added, the down platform lengthened, the central carriage sidings made through roads, and the engine shed moved from the station vicinity. (Exmouth Junction locomotive depot was opened in 1880 and modernized in 1927–8.) But it took complete rebuilding to produce a pleasant, workable station. Renamed Exeter Central, it was officially opened on 1 July 1933, and for a generation was a hive of industry, expresses being split into separate portions for the Plymouth and North Devon lines and their restaurant cars stowed in the sidings, locals continually arriving from Exmouth and other places, and banking engines for ever running down to St David's or panting their way back behind stone trains from Meldon Quarry. But today, charge of the entire station often rests on one solitary railwayman. The two-hourly trains to and from Waterloo treat Central like a wayside halt, and only when the busiest early morning train comes in from Exmouth does the patter of feet recapture something of bygone activity.

THE SOUTH DEVON RAILWAY

> Things have been effected and works done, that even within the life of man would savour of romance, and such have been the monster trains that have as it were flown from here to Teignmouth, and returned from thence to this city . . . On Sunday the officers of the Company did their best for the accommodation of every one, though the number of passengers was such as they could not possibly have anticipated. All, however, was surpassed on Monday. At the station at Exeter,

> from the hour of seven in the morning, it was a most animated scene, the numbers continually increasing in order to the trip down. Indeed, such was the amount that twenty-one carriages were required to contain them, and it is computed that upwards of 1,500 persons went down by the morning train: while about an equal number came up! Language would fail adequately to describe the scene. At Dawlish and Teignmouth bands of music were stationed, flags floated from every tower and eminence, while from the bosom of old ocean came the echo of the joy and hilarity that reigned on land.

Such, as the *Exeter Flying Post* describes it, was the first week-end – Whit week-end 1846 – that Exeter people flocked to the coast. We find it impossible, 127 years later, to realize the thrill of running beside the sea at 40 m.p.h. on those first sunny days of high expectancy on the South Devon Railway. Exhilarating even for third-class travellers, overcrowded in their wooden box trucks. On the other hand, sick at heart must have been the patrons of Teignmouth's high-class lodging-houses when excursionists ended the gentle age of leisure and space they had loved so well – several hundred people on the newly laid-out Den at the same time!

Devon is so associated with holidays that it is easy to imagine it, past and present, almost as a paradise. But in the 1840s housing standards were as disgraceful as elsewhere; unemployment resulted from the decline of the wool industry and mining. The railway was the driving force which quickly swept away the worst of the old order – quickly, that is, after 1852, a year of awakening nationally as well as locally. During its first few years the South Devon had been in financial straits too severe for it to play its rôle properly.

As early as 1836 Brunel surveyed a route, crossing the Teign and Dart and through the South Hams, to Plymouth. But Devon people were not interested and only in 1843, when the GWR, Bristol & Exeter, and Bristol & Gloucester promised liberal subscriptions, was Brunel's new project brought forward – the Plymouth, Devonport & Exeter, quickly renamed the South Devon Railway, whose Bill (a rival project lapsing) had an easy passage, receiving Royal Assent on 4 July 1844. The GWR, Bristol & Exeter, and Bristol & Gloucester jointly provided

£400,000 of the £1,100,000 authorized capital and eleven of the twenty-one directors.

Use of atmospheric power was recommended in the report unanimously adopted at the Company's first meeting, at Plymouth on 28 August 1844. The patent for the atmospheric system had been obtained by Messrs Cleg and Samuda in 1839. The basic idea was simple. Stationary engines pumped out air from a continuous iron pipe laid between the rails. At the top, the pipe had a slot, hinged permanently to one side but opened on the other by a framework of iron plates fitted to the leading vehicle of a train. This vehicle, the piston carriage, had a rod connecting it with a cylinder and close-fitting piston travelling inside the pipe. Air was let in behind the piston, and its force rushing towards the vacuum in front propelled the train. The pipe was then resealed with pressure from a small roller, and a fresh vacuum was then created by the pumping houses.

George Stephenson called it the 'great humbug'; Brunel was confident, brushing aside 'mere mechanical details' and assuming that stationary power of this kind must be cheaper than using locomotives. He estimated a saving of £207,000 in making the line single instead of double: 'The system admits of the working with a single line without danger of collision, certainly with less than upon a double locomotive line.' And such was his faith that gradients and curves were planned – and built – sharper than necessary. Instead of spreading the climbing as evenly as possible between Newton Abbot and Plympton, he grouped it into four 'planes', two facing each way, where larger atmospheric pipes and more pumping engines were to be provided. Taking atmospheric equipment and locomotive costs into account, he believed there could be an initial saving of £67,000 and a yearly saving of £8,000.

Work on the railway started quickly, trains to Teignmouth being promised for July 1845. But easterly winter gales delayed the building of the river and sea walls beyond Powderham, and the opening to Teignmouth was not until 30 May 1846, and to Newton Abbot on 30 December. Locomotives had to be hired from the GWR to work the trains (seven passenger and one goods a day).

By now the Croydon Railway had become atmospheric and

Page 69 Most minor railways have long been closed but interest in them lingers on

Plate (13) *The Clevedon 'headquarters' of the Weston, Clevedon & Portishead Light Railway.*

Plate (14) *A Bideford and Westward Ho! train with American-type coaches and cow-catcher on the locomotive at Bideford Quay.*

Plate (15) *Desolation—note signal-cabin and posts—on the Lee Moor Tramway.*

Page 70 Branch line miscellany

Plate (16) One of the former LSWR *Adams 4–4–2 tank locomotives which was considerably older than the Lyme Regis branch it served for many years.*

Plate (17) 'Queen', the engine which helped to make Brixham prosperous but had to be mortgaged by its owner.

Plate (18) At Uffculme on the Culm Valley line, c. 1880.

Page 71 Dartmoor

Plate (19) Moretonhampstead was typical of the West Country's wooden shed stations.

Plate (20) Two engines were needed to haul the six-coach train on the Princetown branch's last day,

Plate (21) . . . but there were no passengers waiting in the fog at lonely King Tor Halt.

Page 72 A page of narrow gauge

Plate (22) Crossing the Exmoor foothills on the Lynton and Barnstaple section of the Southern Railway.

Plate (23) 'Quintus' resting in the woods near Creech Grange on the Furzebrook clay line in the Isle of Purbeck.

Plate (24) A Penlee stone quarry train returning empty from Newlyn harbour half a century ago.

Brunel told the directors he wanted to gain from its experience before proceeding too far: 'I deliberately repeat my opinion, that I entertain no doubt whatever that you will be perfectly satisfied with the result.' But a year later, we find from the recently discovered journal of P. G. Margary (then a young assistant, later engineer of the South Devon, then of the Cornwall Railway and finally GWR Plymouth divisional engineer – a man always much respected), almost everything had still to be learned.

In January and February 1847 Margary gives instructions for the engine houses to be connected with the telegraph; he engages a policeman at 18s a week, including Sundays, and dismisses a workman 'as he made use of very impudent language to me which I could not allow'; above all, he tells the staff they must hurry, hurry, hurry with the engine houses and their equipment. On 25 February the first piston carriage came down by a Bristol & Exeter train. Enthusiasm was high:

> Started at 6.0 p.m. for Turf, towing a locomotive behind. Went on very slowly to Turf, there being a large quantity of water and dirt in the pipe.
>
> 11 March: at 6 p.m. we started with a train for Turf but a large quantity of water having got into the pumps, the engine [*sic*] were not able to work as fast as they should have done.
>
> 1 April: Went after the 12.5 by the atmospheric train to Star Cross and returned after the 2.20 had passed. Very bad results.

And on 28 August Brunel reported that the difficulties were 'beyond what any previous experience would have justified anybody in anticipating'. Experimental trains were now running regularly, but the equipment – not least the large stationary steam engines – still gave trouble and the directors agreed to a Midland Railway (successor to the Bristol & Gloucester) request not to incur expense on the system beyond Totnes.

On 13 September 1847 atmospheric trains began carrying passengers to Teignmouth, and on 10 January 1848 to Newton Abbot. 'I think we are in a fair way of shortly overcoming the mechanical difficulties,' said Brunel. Pipes were then being laid between Newton Abbot and Totnes, which had been opened for passengers on 20 July 1847. They were never used: from 9 September locomotives took over the working of all South Devon trains. Brunel had come full circle and recommended the atmospheric's abolition. £426,368 had been spent on the system; only £50,000 was later recovered by the sale of materials. More serious in the long run was the legacy of the single line and the steep gradients.

Between stations atmospheric trains reached high speeds and were popular for their smoothness and cleanliness. But stops had to be made at all stations as the piston automatically left the pipe, and breakdowns became frequent. The worst troubles concerned the longitudinal valve, and have been summarized thus:

> The natural oil of the leather of the longitudinal valve was sucked out of it by the partial vacuum within, rendering it pervious to air and water: the leather then soon became either saturated with water, which soaked through it in wet weather or froze it in cold, rendering it too stiff to shut down, or so dry and hard that it tore easily. The remedy adopted—an immediate application of seal-oil—often could not be made in time, and had to be frequently renewed and was expensive.

And later:

> Under the joint action of water on the leather, and of the affinity of iron for tannin—and on the enduring presence of tannin within its texture the consistency of leather depends—a destructive decomposition had long been at work; the oxide, established on the iron plates of the valve by continued contact with damp leather, had been steadily abstracting the tannin; thus the leather had become converted into an ink-stained and comparatively decomposed tissue. Large portions of it became torn, and incurably pervious to air.

Totnes to Laira, on Plymouth's outskirts, was opened on 5 May 1848, and on 18 December that year a five-mile branch (authorized 1846), independent of the main line from its separate shed station at Newton Abbot, was opened to Torre.

From Exeter to Newton Abbot the South Devon is almost level, first across Exminster marshes, in sight of the Exeter Ship Canal, then down the Exe (lake-like at high tide, fascinating for bird life at low), along the sea between Dawlish Warren and Teignmouth, and then up the Teign estuary. One of the two remaining atmospheric pumping-houses is beside the track at Starcross: it now serves as a chapel. (The other, at Torre, on the opposite side of the road from the fire station, never came into use.)

In its early years the sea wall caused great worry and expense. In December 1852, for instance, much rock and soil fell at Breeches Cliff, east of Parson's Tunnel, the longest of the five between Dawlish and Teignmouth. The sea made its first inroad in 1854: 50 ft of the wall near Teignmouth splashed into the waves. Traffic was interrupted for about a fortnight – one of the least inconveniences of that bitter Crimean winter. Since 1873 damage has been superficial, such as the washing away of ballast, though few winters pass without some interruption to the traffic, usually the closing of the down track for a few hours at high tide.

Climbing begins immediately after Newton Abbot: the final section of Dainton incline up to Dainton Tunnel is at 1 in 37. After the tunnel comes the sharp, curved drop to Totnes, where the placid River Dart is crossed at sea-level. So climbing has to start again to cross the southern flanks of Dartmoor: Rattery incline (as Hemerdon incline) was built double. The rest of the South Devon was originally single. After Rattery comes Marley Tunnel, 869 yds, needed only because Sir Walter Carew of Haccombe would not have the railway spoiling his view. The line's summit is near Wrangaton, but it is at Bittaford, with hut circles, tumuli and other prehistoric relics but a mile from the track, that the moor's wildness is nearest. Just before Ivybridge and on either side of Cornwood valleys, deep cut by streams racing seaward, are crossed by handsome stone viaducts which replaced Brunel's timber ones. The latter, all over 220 yds long and 100 ft high, were frail-looking, their piers formed of two slight, tapering pillars, with occasional struts between them. Designed for atmospheric trains, they often vibrated alarmingly, though strengthened for locomotives, when expresses thundered across. On Brunel's later timber viaducts, as in Cornwall, the stone pillars stopped well below the trestles, a network of large supports radiating out from them. With these first West Country examples, however, the piers were taken right up to rail level. Hemerdon incline – 1 in 42 for two miles – begins a mile after Cornwood.

At first the South Devon was generous with its train service and provision for poor travellers, but as early as August 1847 it was reported:

> Your Directors had been desirous to hold out every encouragement and facility to Third Class Passengers that might not interfere prejudicially with the amount of First or Second Class traffic, but the extent to which advantage of the accommodation thus afforded was taken by classes of persons for whom it was not intended, has obliged your Directors, in opening a further portion of the line to Totnes, to reduce the number of Third Class Trains, and so to regulate the times at which they run as to adapt them, as far as circumstances would admit, more exclusively to the wants of the labouring classes.

This action merely contributed to the loss of nearly £3,000 during the first half of 1848: the directors had just heard of it when they decided to abandon the atmospheric, Thomas Gill, the chairman, being alone in urging further trials. He resigned, and after rebuff at a public meeting at Exeter on 6 January 1849 left the Board. The capital was now all but £2,000,000 and between April 1848 and August 1851 no interest was paid on preference shares; goods traffic was slight, bringing in little more than £1,000 a month – only in 1852 was a connection established with Teignmouth Quay – and an effort to swell passenger receipts by increasing fares and cutting the timetable failed. At one time the Company's survival seemed doubtful, but in 1851 the preference shareholders agreed to commute the £25 shares carrying 30s interest, guaranteed till 1857, for a perpetual 10s 9d. That allowed a 1 per cent dividend on ordinary shares, which mite so relieved the owners that they presented the new chairman, Thomas Woollcombe, town clerk of Devonport, with £500 worth of shares.

As already seen, 1852 marked the beginning of better times. Fares were much reduced; and then came the excursion trains! Too long for any platform, they carried 1,000 people or more, usually spending as much time at stations as in their rockety motion. Almost invariably something went amiss, 'dispatched with tolerable accuracy' being the highest praise. It was in July that the first West Country excursion ran for Londoners – 1,500 of them in 33 carriages – allowing them a week in a choice of resorts. The fare to Plymouth was £1; the journey time 13½ hours. Unfortunately we have no details of locomotives on these auspicious occasions.

The breach caused by the atmospheric failure was filled at first by the GWR, but from 1851 Geach, Evans & Co. supplied locomotives for £300 minimum a week or so much a mile – 1s 3d, for instance, with 12 coaches east or 6 west of Newton Abbot. Though this was less than the GWR's charge, the firm made a big profit; premiums were paid to drivers for fuel economy and the speed of goods trains climbing the banks was restricted to 6 m.p.h. Until 1859 there were 16 engines, four for goods and the rest 'Comets', four-coupled bogie saddle tanks, nicknamed 'Hunchbacks'. The crew had no cab, and as on most later South Devon engines, in wet weather the fireman scrambled on to the tanks above the boiler to pour sand down a funnel, whence it trickled to the slippery rails. The 'Comets' did excellent work. In 1859 the contract was renewed on terms more favourable to the South Devon, and 12 new bogie saddle tanks, the 'Hawk' class, were added. The South Devon operated the independently owned branches (including the Torbay & Dartmouth), and also provided power for the Cornwall Joint Committee and later the West Cornwall Joint Committee too, so Newton Abbot assumed its rôle of chief locomotive depot in the South West.

To complete the locomotive picture, Messrs Geach, Evans's contract ended in 1866, when the South Devon bought the 40 tank engines then at work. By 1876, 85 locomotives were employed on the broad gauge and its branches between Exeter and Penzance: livery was dark green, picked out in black. *Tiny*, an untypical specimen, is the only surviving broad-gauge locomotive. Built for Plymouth's Sutton Harbour branch, it later went to Newton Abbot and was rescued after years of disuse in the workshops, where for a time it had toiled as a stationary engine. It now stands on the down platform. The South Devon's coaches, uncompromisingly rectangular, and with square windows, were brown.

The Company never became progressive. Even after 1852, many of the timetable alterations were for the worse. Torquay's (actually Torre's) service was cut in 1855, and as the next season was beginning it was cut again, most evening trains disappearing. 'Surely the directors will not be so blind to their own interest as to allow this crying evil to continue,' said the mystified local paper.

That the directors did not always know their own business is suggested by increased steamship traffic at this time.

Meanwhile ships were being used on the Dart as feeders to trains at Totnes. The service was sponsored by Charles Seale Hayne (founder of the agricultural college), who was also 'the indefatigable champion' of the proposed Dartmouth & Torbay, incorporated in 1857 to extend the Torre branch. Capital coming slowly, only the three miles from Torre to Paignton were first tackled, and even they (with 20 bridges, a viaduct, and a tunnel – opened out in 1910) proved difficult enough. Opening date was 2 August 1859. Three more miles to Churston, known as Brixham Road till the Brixham branch was opened, followed on 14 March 1861. Further delay was occasioned by attempting to take the line to Dartmouth itself. The Lords rejected the proposal, so the original scheme for a terminus at Kingswear and steam ferry over to Dartmouth was revived. Work started in 1862, and the 3¼ miles, involving heavy engineering including 495-yd Greenway Tunnel, were opened on 16 August 1864. Since then Dartmouth has been noted for its railway station without trains.

At this time the South Devon itself was concentrating on doubling the main line, beginning in 1855 with Newton–Totnes, taking advantage of the Torquay branch for the first mile and establishing Aller Junction. After this, work was usually afoot, the easiest sections being chosen first. By the end of the Company's independent existence, all the main line except Dawlish–Teignmouth and Rattery–Hemerdon had been completed. A second platform was added at many intermediate stations; in 1861 Newton Abbot's three sheds – for up, down and Torquay trains – were replaced by one building. Traffic continued to grow and in 1874 a third track to Aller was laid, again giving the branch separate access to Newton station. The arrival of a train at Torquay now brought a healthy bustle as the wealthy changed into their carriages and servants handled the piles of luggage. Till the end the South Devon snubbed the 'lower orders': on the main line they could use only three of the ten down, and nine up, daily trains.

The end came suddenly, the GWR being approached as soon as it had taken over the Bristol & Exeter. Terms were £65 of GWR ordinary stock, to be increased to £70 after seven years, per £100 of South Devon. Some shareholders complained that the directors had panicked because of coming LSWR competition at Plymouth, but on hearing that continued independence would entail raising £1,000,000 (for gauge conversion, doubling and viaduct rebuilding) only 213 votes were cast against the merger, compared with 9,930 in favour, at the meeting of 17 December 1875. The GWR worked the line from 1 February 1876, the South Devon Company being dissolved on 1 August 1878. Its dividend record was poor. Not until 1864 did the figure reach 2 per cent; in 1872 4 per cent was paid, but a decline to less than 2 per cent followed.

THE MAIN LINE AFTER 1876

Between 1841 and 1871 Torquay's population had increased from 5,982 to 21,657 and the number of houses from 926 to 3,071 – the well-planned terraces and villas we know today were built. Ever since the Royal Hotel had been reconstructed in 1828, being adapted 'for the reception of families of the first distinction', the 'Montpelier of England' had concentrated on luring the rich. Hitherto its success had been astonishing; but now holidays abroad became fashionable. The old clientèle shrank, and as yet few middle-class visitors replaced it. At this time, of course, Torquay's chief fame was as a winter resort. Paignton, a 'neat and improving village', when the railway came had a population of 3,700 in 1855 and was catering for proportionately more summer visitors than Torquay; but its growth, too, was temporarily slowed.

A new wave of development began in 1890. As an appreciable number of middle- and lower-class people began taking an annual fortnight by the sea, the emphasis changed gradually from quality to quantity and from winter to summer. Train services were considerably improved in 1890 for the first time since the GWR took over: see Chapter XI. Excursion traffic too now developed rapidly. Teignmouth was generally slow in adapting to changing times, but being on the main line it quickly developed a substantial business in catering for Sunday School treats and other trippers. In 1905 a rail motor service was introduced to augment the trains between Exeter St Thomas and Teignmouth, and a station was opened at Dawlish Warren. Excursion tickets for football fans were now issued weekly. Goods traffic also vastly increased during and after 1892: consignments arrived by train which so far had remained loyal to boats, or had perhaps been sent by narrow-gauge rail to Exeter and collected by road.

The GWR was slow to continue doubling the main line where the South Devon had left it. Only in 1884 was the next section tackled – the sea wall from Parson's Tunnel to Teignmouth. Teignmouth station was rebuilt and a short tunnel outside it was opened. Nothing more was done until after the gauge conversion, Rattery to Hemerdon being completed in 1893. Throughout that year a large labour force was employed in the Ivybridge district building the new viaduct. A second single-line Marley Tunnel had been started in 1891, causing a landslip which blocked the original bore between 30 May and 4 June that year. (1891 is better remembered by old Devonians for the great March blizzard, when the services were totally dislocated and an express derailed by a snowdrift spent several days at Brent.) That left only the Dawlish–Parson's Tunnel section: the first of the five tunnels through the red sandstone cliffs was started in 1902, traffic not being interrupted as the liberal broad-gauge clearance allowed trains to run through a movable arched steel frame while labourers worked on top of it. Not till 1 October 1905 was there a continuous double line from Exeter to Plymouth – a double line that would have been ready fifty-six years before but for the 'atmospheric caper', as Devonians like to call it. The Torquay branch was doubled to Kingskerswell in 1876, Torquay in 1882 and Paignton in 1905. Other pre-1914 improvements included alterations to Kingswear jetty and quay in 1904 and, back on the main line, the opening of Hackney goods loop and marshalling-yard east of Newton Abbot in 1911.

Between the wars the biggest scheme was the building of the present Newton Abbot station in 1927, at a cost of £140,000. Of the nine platforms, one is detached from the station proper and served the Moretonhampstead trains; two long islands provide the others, Nos. 1 and 2, for instance, being different ends of the same face, with a scissors crossover in the middle (connected with one of the through lines placed on either side of the station) to help in dividing trains. Nearly all expresses have portions for Plymouth and for Torbay. Newton was gaily decorated on 11 April 1927, when GWR officials accepted the clocks presented by the enthusiastic townspeople. By this time, incidentally, Newton's population was 15,000, compared with only 6,000 in 1871.

In 1935 a new double line, eight and a half miles long including a tunnel, was proposed inland between Dawlish Warren and Newton Abbot (or, to be precise, Bishopsteignton). This was part of a Government scheme to help the railways and industry. Traffic was growing and it was felt that the whole route between Exeter and Newton Abbot justified quadrupling, but there was not room for four rails along the sea wall. Various preparations had been made when war came again. The land was resold in the 1950s: not only would the cost of the work be prohibitive today, but for several reasons better use can now be made of the double track. Incidentally, on the South Devon line up and down running loops are available to passenger trains at Exminster, Dawlish Warren, Totnes, Brent and Hemerdon, and the track between Newton Abbot and Aller Junction is quadrupled. Before the war the Torbay branch was doubled from Paignton to Goodrington, where the Sands Halt was opened in 1928 and numerous passenger sidings built. 1956 saw more sidings and a turntable, the Halt rebuilt, and the level-crossing abolished. Unfortunately the level-crossings on either side of Paignton station remain.

By 1958 the number of visitors to South Devon had trebled compared with pre-war days, and a greater proportion of them were coming in the season's sharp peak. In 1956, for example, a total of 98,042 long-distance passengers (i.e. from at least as far away as Bristol) alighted at Torre and Torquay on the fourteen Saturdays from 16 June to 15 September inclusive. Well over half of them – 51,057 – came on just five Saturdays and no fewer than 14,594 on the single busiest day, 28 July. Of these, 4,348 came from Paddington, 1,116 from the Bristol area, 2,290 from the North (*via* the Severn Tunnel), 2,676 from the London Midland Region *via* Bristol, 1,171 from South Wales, 647 from the Eastern Region *via* Didcot, and 2,296 from the Western Region's North Warwick line from Wolverhampton and Birmingham.

1956 marked the zenith of this expensively dense seasonal traffic, most of which was carried on Saturdays, and by 1962 there was already a sizeable reduction in the number of long-distance arrivals on the fourteen busiest days:

	1956	1962
Torquay and Torre	98,042	76,252
Paignton	70,365	59,928
Teignmouth	14,550	14,448

From the railway's point of view there was the relieving feature that the drop was especially sharp at the season's peak. Thus the number of long-distance arrivals at Torquay and Torre on the five busiest Saturdays was down from the 51,057 already quoted for 1956, to 32,887 for 1962, and the busiest single Saturday of all showed a drop from 14,594 for 1956, to 9,059 for 1962. For it was only at this point that staggered holidays, advocated for so long, actually began to happen; and they happened because the increasing proportion of people coming by car (hitherto the railway had carried more people each year, but still a declining proportion of the total) could not tolerate the traffic jams that clogged up the West Country's inadequate road system at the busiest weekends. It was, alas, the railway's ability to carry the crowds it did (albeit in some discomfort) that delayed the extension of the season; that in turn encouraged the growth of shoddy accommodation, including camps eating into the beauty of the countryside, to cope with peak demands, and discouraged investment in more luxurious hotels needing longer occupancy to be economic.

Since 1962 the railway's share has declined a lot further but has now stabilized at about a fifth of the total. Few special facilities are now provided for the peak Saturday business, yet punctuality is better than ever enjoyed in steam days and rarely do passengers have to stand. The railways have also benefited from the increased holiday activity in May and June, and again in September and October; though no new hotels have been built, since around 1958 many existing ones have been extended or completely remodelled and most of them keep busy from Easter to the end of October. Thus many people taking short second and third holidays fill seats on the increased number of ordinary long-distance trains for over half the year, and even in late summer and winter Londoners coming to spend the weekend at, say, the

Imperial Hotel at Torquay travel alongside Devon young men and women returning home for a night or two from their up-country jobs. The railways were slow to realize the potential of this week-end business and for some years the Sunday service from South Devon to Paddington was disgraceful; now a series of good expresses carry people back in comfort.

Most of the investment in hotels has taken place in Torbay where there has also been big residential development, especially in Paignton and Brixham, though also in some villages inland. In 1921 the three Torbay towns (Torquay, Paignton and Brixham) had a combined population of 61,053; in 1931 of 72,911; and in 1961 of 94,883. Now with over 100,000, Torbay enjoys County Borough status until, along with Exeter and Plymouth, it will be absorbed into an enlarged Devon under the new local government reorganization. Teignmouth expanded slowly after the war, picked up dramatically at the end of the 1950s, and residential development continues apace, as it does also in the Dawlish Warren and Newton Abbot areas. Newton Abbot attracted valuable light industry after the war, but unfortunately before a run-down of the sizeable railway facilities, so that there has been nothing to take their place and unemployment is high. Once the locomotive headquarters of the South West, Newton Abbot is fast becoming little more than a wayside station. Even the expensive diesel maintenance depot, the pride of the Western Region, has been closed, along with the carriage and waggon repair shops, which like the offices over the booking hall are now occupied by David & Charles, the publishers. But Newton does fulfil an important holiday role as a motor-rail terminal for services to a variety of destinations, Stirling perhaps being the most popular.

The first edition of this book continued comment on the holiday trade by saying that many holidaymakers bought 'run-about' tickets, while other South Devon excursion traffic was such that up to 8,000 Exeter people travelled to Dawlish Warren alone on summer Sundays. Times have changed since 1958. Though 'runabout' tickets are still available, and a handful of passengers still go from Exeter to Dawlish Warren by train on sunny Sundays, and there is the remnant of a commuter business from Torbay, Newton Abbot, Teignmouth and Dawlish to Exeter, local traffic is now of little importance anywhere in South Devon. Even in high summer it comes as a surprise to see holiday families getting on a train at one station and alighting at the next. In 1958, in winter, about 20 daily expresses went westward from Exeter, plus 12 locals. In 1973, the expresses number 27, but the local trains only five. Though all stations between Exeter and Paignton remain open apart from Exminster and Kingskerswell, South Devon has virtually abandoned the railways except for long-distance journeys. However, it is only fair to stress that were it not for the improved express trains that can take them quickly back to London should the need arise, many retired people would have been afraid to cut themselves off so far in the West. Trains are still the quickest as well as most reliable way of reaching London, and to that extent South Devon's building boom directly depends on the railway.

Even most of the few stopping trains that remain carry few passengers; and the local bus services are not doing well either. The GWR itself was a road pioneer, but in 1929 sold its West Country services to the Western National Omnibus Company, which in 1933 sold those in the Exeter and Torbay districts to the Devon General Omnibus Company, not so closely associated with the railway. There has never been much rail–road co-operation, to the detriment of public transport generally. Considering the size of its population, South Devon is surprisingly un-public transport minded.

The bulk of the main line's freight traffic is on its way from Plymouth and Cornwall, but there have been big recent changes here, too. In 1958, an annual 250,000 tons of coal were going down the line and 150,000 tons of clay coming up. The clay traffic continues, much of it carried by a daily clay 'liner' train to the Potteries – the only freight liner using the route since a London–Plymouth–Cornwall service had to be abandoned for lack of support. The coal, however, has largely disappeared, since Exeter's coal concentration depot at Exmouth Junction is now railhead for supplies throughout Devon and Cornwall, right down to Penzance. Newton Abbot's Hackney marshalling yard is now little used, marshalling being concentrated at an extended Riverside Yard at Exeter, and loads are kept light to avoid using banking engines over the inclines between Newton Abbot and Plymouth. Teignmouth's harbour no longer has a rail link, though initially the harbour developed because of its main-line position and for many years coal and timber were distributed locally by train. To look at one station's freight traffic as it was in 1958, Totnes sent outwards yearly 28,000 parcels, including the products of a bacon factory, 600 tanks of milk for London from the local creamery, 6,000 tons of mixed cargoes including locally-produced sweets and timber imported by ships which still use the Dart, and 100 trucks of cattle. Of the 20,000 tons of inward traffic, a quarter was feeding-stuffs and fertilizers. Only the milk remains, since the handling of all other freights is concentrated at Newton Abbot – and at a goods yard much reduced at that, part of it being devoted to a coal depot, the coal arriving from the Exeter concentration depot by road. Cattle traffic has ceased on British Railways as a whole. But it was only in July 1962 that a siding was installed at the new Totnes cattle market; it was quickly threatened with closure, reprieved with a flurry of publicity, and then closed.

Totnes is the only intermediate station now open between Newton Abbot and Plymouth. The five between Brent and Plymouth were closed on 2 March 1959 and Brent on 5 October 1964. Plymouth's multiple-aspect signalling system is currently being extended eastwards to Totnes.

In 1958 about 30 winter trains travelled down the Torbay line to Paignton, 20 of them continuing to Kingswear. Only 20 now reach Paignton, and many carry very light loads, albeit most passengers are at the beginning or end of long-distance journeys. Since October 1972 Paignton (carriage sidings are still maintained at Goodrington) has been the end of the line for British Railways, the Dart Valley having taken over the line on to Kingswear as a second string to its Totnes–Buckfastleigh venture. Dartmouth was a town in the doldrums both before and after the war, losing 1,400 people since 1921 through failing to develop tourism sufficiently to offset the disappearance of the coal bunkering and other trades. But many trains, including through ones from Paddington, continued to carry useful numbers of passengers through to Kingswear, and the decision to treat Paignton–Kingswear as a self-contained branch, in practice given poor connections with expresses, was a pity. Cadets attending Dartmouth

Naval College were then met off the train at Newton Abbot, and the absence of through trains (a daily summer one from Plymouth had remained busy) killed traffic at the rebuilt Goodrington Sands Halt. Earlier there had been a threat to close Dartmouth's station without trains when the steamship *Mew* (1908–54) had to be withdrawn from the ferry service because of old age, but eventually two new diesel ferries were built. Kingswear once handled important coal traffic, much of it for the former gas-works between Torquay and Paignton.

Unlike other lines run as nostalgic attractions with steam power, the Paignton–Kingswear section of the Dart Valley has a winter service of two morning and two evening trains each way on Mondays to Fridays, carrying among others children attending the school at Churston built after the war beside the station where three trains (from Kingswear, Brixham and Paignton) arrived simultaneously each morning. The Dart Valley has to stagger arrivals since the crossing loop was taken out some years ago; its restoration is but one of the tasks ahead for voluntary labour.

BRANCH AND MINOR LINES

In 1790, James Templer of Teigngrace began the Stover Canal to carry ball clay (for notes on the clay industry see page 148) from the Teigngrace area to the Teign estuary. The HAYTOR tramway, a feeder to the canal and the first and oddest of Devon's railways, was opened on 16 September 1820.

> On Saturday Mr Templer, of Stover House, gave a grand fête champêtre on Haytor Down, on the completion of the granite rail road. The company assembled at its foot on Bovey Heathfield, and in procession passed over it to the rock. A long string of carriages, filled with elegant and beautiful females, multitudes of horsemen, the workmen on foot, the waggons covered with laurels and waving streamers, formed in their windings through the valley, an attractive scene to spectators on the adjacent hill. Old Haytor seemed alive: its sides were lined with groups of persons, and on its top a proud flag fluttered in the wind.

The peculiarity of the line was its being made of the material it was built to carry – granite. Hewn blocks formed the flanged rails, varying from 4 to 8 ft long; the gauge was 4 ft 3 in. The iron waggons' wheels were flangeless. Built without Act of Parliament, the tramway climbed nearly 1,200 ft in six miles without a steep gradient requiring extra horses. This was the personal achievement of the resourceful George Templer, son of James. Trains were usually of several flat-topped trucks. The cost of the granite was high because at Teigngrace it had to be transferred from the railway to the canal, and at Teignmouth from the barges to sea-going ships. But for many years it was in good demand, and the quarrymen broke the loneliness of the moor. Then cheaper granite came on to the market, and in 1858 the tramway fell into disuse, though periodically for many years the quarries were worked for specially long pieces. Haytor village melted away, as have so many Dartmoor mining settlements in various eras. Much of the track remains intact and almost all the layout of the upper terminus – six quarries were served – can be traced. The visitor will enjoy following the granite slabs, as indigenous as the scattered boulders round them. The most fascinating course is to strike past the Tor itself until the first 'rails' are found and then follow the lowest, or most northerly line, from the windswept down to old Holwell quarry, with its imposing smooth face, a few bushes struggling for existence on a crevice. Here, interrupted only by the cry of the curlew, you may commune with the spirits of the men who cut granite for parts of the National Gallery and London Bridge; and keen though you may be on railways, you will not regret that a 1905 scheme for electrifying the line and running a frequent service of trams for visitors failed to mature. The tramway is now protected as an ancient monument.

A section of the route of the granite tramway, between Teigngrace and Bovey Tracy, was used by the MORETONHAMPSTEAD & South Devon Railway. Opened on 4 July 1866, this line was worked by the South Devon which, however, contributed only £500 of the £105,000 allowed capital. Local people were slow to subscribe and bad weather delayed construction; but there were no major engineering works for the broad-gauge line, 12 miles long and climbing 550 ft from the junction at Newton Abbot. The initial train service was poor, the South Devon seeming to bear the small Company a grudge. Traffic was sparse till the end of the century, when well-filled trains of six or seven six-wheeled coaches became familiar at the Moretonhampstead terminus, with its overall wooden roof and mean exterior, which remains today. Then, in 1906, the GWR began running buses to Chagford in connection with the trains. As the Torbay resorts grew, visitors flocked to the railway: Lustleigh's stationmaster was tourist-conscious enough to keep a visitors' book in the waiting-room. Despite road competition the branch was busy until the war, but no effort was made to reattract business afterwards, and with traffic run down as the result of an inconvenient service, passengers ceased to be carried on 2 March 1959. A recommendation from the Central Transport Users' Consultative Committee that the position should be reconsidered when diesel railcars became available in the area in 1961 was ignored. For a few years after the war lignite (bituminous coal) traffic flourished at Heathfield; the mine is now defunct, and only a little clay goes by rail from Teigngrace. The section above Bovey was closed to freight in 1964, and the rails were withdrawn from Bovey itself after a special one-day service of passenger trains from Newton in 1971. But Heathfield has sidings serving a banana-ripening factory and oil-distribution depot, and a little ball clay comes from Teigngrace.

Numerous and bold were the plans for an inland railway between Exeter and Newton Abbot, but eventually it took two Companies to complete a single line. The Teign Valley Railway needed nine Acts to coax it into existence and another three afterwards – the record for a line of seven and three-quarter miles. At one time the LSWR was interested, and once the Company found itself in Chancery, but it was eventually built under GWR protection, and worked by the GWR when opened from Heathfield (on the Moretonhampstead line) to Ashton on 9 October 1882. Soon goods went on another mile. Until the Moretonhampstead's gauge was converted, the Teign Valley was an isolated standard-gauge concern with a side tank locomotive, several six-wheeled carriages and an *esprit de corps* all its own. The EXETER Railway sanctioned in 1883 to finish the route to St Thomas, Exeter, was saddled with a hopelessly large capital. Engineering works included two tunnels and numerous cuttings – 600,000 cu. yds of earth had to be removed – largely necessary only to placate landowners. Opening was not until 1 July 1903. The complete St Thomas–Heathfield run was generally known as the Teign Valley and was worked by the GWR as one unit, though some short-distance locals were run. Twice a week Chudleigh Knighton Halt was the start for a trip to Newton Abbot: such was the importance of the railway before buses and cars. Once the branch was a useful alternative to the main line, but since nationalization Plymouth traffic has taken the Southern's Okehampton route if, for instance, the Dawlish sea wall is closed. The emergency valve of the single line with its 1 in 56 gradients thus reduced, and with mineral traffic almost gone and passengers scarce, Exeter–Christow was closed completely from 9 June 1958. Goods traffic continued between Heathfield and Christow until the track was damaged by floods between Trusham and Ashton in September 1960, and now the last section of the branch has been lifted following the end of traffic from Trusham concrete works.

The Torbay & BRIXHAM Railway, opened on 28 February 1868, was the result solely of R. W. Wolston's enthusiasm for helping his fishing-port. He took 1,770 of the 1,800 £10 shares, completed the two miles of broad gauge when the contractor defaulted, and possessed his own small 0-4-0 engine, *Queen*; but he let the South Devon work the line as his agent. Every account showed Wolston in the red – till he was forced to mortgage his engine. Only when friends engaged an accountant was it discovered that the South Devon was allowing him no commission on traffic going beyond Churston. A case was taken before the Railway Commissioners in 1876, but then Brixham trains were charged ridiculous fees for the use of Churston station. In desperation Wolston considered becoming narrow gauge and quite independent; but another appeal to the Commissioners resulted in his receiving £2,000. That gave the Torbay and Brixham confidence to run the line itself, but in 1883 it was sold to the GWR for £12,000. Brixham station was inconveniently situated above the town and passenger traffic was thin, but Wolston would be pleased to know that the port escaped the decline of most in the West, fish leaving by train almost every day until the railway's closure on 13 May 1963.

From the traffic point of view the obvious route to ASHBURTON was from Newton Abbot, but a century ago railways had a monopoly and could afford long deviations to take advantage of river valleys. Actually, the first proposal was for a line from Newton Abbot; but nothing happened till the Buckfastleigh, Totnes & South Devon Railway was incorporated in 1864, gaining powers the next year for the extension to Ashburton. Opening was not until 1 May 1872. By then Ashburton had lost much trade and population, but about half of Devon's remaining wool trade was to be found in the Dart Valley and as late as 1890 Buckfastleigh provided more traffic than even Newton Abbot. In the motor age the branch quickly became a white elephant, and it was closed to passengers from November 1958, and to goods from 10 September 1962. The track remained however, and after prolonged negotiations in April 1969 the Dart Valley Railway with Great Western locomotives and rolling stock began a tourist service from Totnes (at a point called Riverside where a loop has been installed just short of the main line) to Buckfastleigh. It was hoped to continue through to Ashburton but the Devon County Council, having opposed the line's closure, now opposed its re-opening, and after a public enquiry use of the track bed for improvements to the A38 was permitted. Through trains ran from London and South Wales to Ashburton on a final day of celebrations in October 1971. Since then Ashburton's terminus, still retaining its all-over wooden roof, has been an isolated Dart Valley exhibit, while with the truncation of the line immediately beyond Buckfastleigh a new layout has had to be provided there and a loop has also been installed at Stoverton. It is intended to keep Totnes–Buckfastleigh alive in addition to the Paignton–Kingswear section.

The TOTNES QUAY goods branch was opened in 1873; its Act forbade the use of 'Atmospheric Agency'. For most of its life it was worked partially by horse power, but locomotives eventually went over the level crossing to reach the quay. Closure came in 1965.

South Hams people were even less enthusiastic than their neighbours about supporting a railway, many positively arguing that the self-contained district would be better without trains. A line was started in the 1860s but was abandoned for want of capital, and it was only on 19 December 1893 that the GWR's KINGSBRIDGE branch was opened. With three good coach services and a lively harbour, Kingsbridge may not have missed trains as much as other places would. By 1935 most local passengers had taken to the road, but the line shared in the rush on summer Saturdays, when up to 1,000 visitors were carried in each direction; freight remained substantial until total closure on 16 September 1963.

The REDLAKE tramway, serving, and owned by, a china-clay quarry in the middle of the wild moor, five miles from the nearest road, was of 3 ft gauge, 7½ miles long, rose from 400 ft above sea-level to 1,400, and was opened on 11 September 1911. It started with a cable-operated incline from a siding on the GWR near Ivybridge. Above the incline, locomotives were used, first on the scene being 0-4-2 *Handson*; in the morning it took up workmen and trucks of coal for the pumping engine; in the evening it returned with men and sand. The clay itself was piped down. Apart from one break, traffic was steady until the clay company failed in 1932. The course of the line, with its ballast still clean, is useful to hikers.

CHAPTER V

North Devon

TO BARNSTAPLE, TORRINGTON AND ILFRACOMBE

The railway made South Devon's fortune and lost North Devon's. Even today the holiday season is shorter and less profitable along the Bristol than the English Channel coast and, lacking industry, the Taw-Torridge district has an unemployment rate considerably above average. Not only did trains arrive late on the scene but, when they did come, they were slow and infrequent and stayed so. A long section of even the chief line, from Exeter to Barnstaple, remained single and without full-length loops at the crossing stations. Had it been doubled even after the 1939–45 war, North Devon's outlook might be brighter. But not surprisingly the district has long lost interest in railways. The Southern system west of Exeter, incidentally, has always been known as 'the withered arm'.

It was not for want of scheming that the steam age invaded the area late. Ignoring most of the fruitless plans, the unhappy story begins in 1838, when the Taw Vale Railway & Dock Company was incorporated to make docks at Fremington and a railway thence to Barnstaple. Nothing happened until in 1845 the Company revived its powers with a view to selling itself to the proposed North Devon Railway, which with GWR, Bristol & Exeter, and South Devon support was to continue the proposed broad-gauge route of the Exeter & Crediton on to Barnstaple. The Exeter & Crediton obtained its Act in 1845. (An earlier Exeter & Crediton, of 1832, come to nothing.) The North Devon line was too late for Parliament that year, and having refused to buy the Taw Vale found itself opposed by a Taw Vale Extension narrow scheme (with LSWR backing) in the 1846 session.

The North Devon failed Standing Orders, but though alone in the field the Extension's outlook was bleak; for one thing, local people, wanting to avoid a break of gauge at Exeter, strongly preferred the broad. The directors therefore pretended to drop LSWR interest, and with broad-gauge support obtained their Act on 7 August. This directed that the gauge should be 'such as the Board of Trade shall in its Discretion approve': without this section, under the 1846 Regulation of Gauge Act the gauge would automatically have been narrow and the line was to be leased to the Bristol & Exeter, an entirely broad-gauge concern. Breach of faith was not suspected until some months later the same Taw Vale directors who had proposed the Bristol & Exeter lease recommended rejection. Meanwhile, it had also been understood that the Exeter & Crediton would be leased to the Bristol & Exeter, but when the formal agreement was presented in January 1847, the majority of the small Company's shareholders rejected it.

The LSWR had bought shares in both the Taw Vale Extension and the Exeter & Crediton, in the first case with the directors' compliance but in the latter without their knowledge. The broad double-line Exeter & Crediton, leaving the Bristol & Exeter at Cowley Bridge, over a mile from Exeter, was now nearly ready, but when the directors prepared to open it they were stopped by the LSWR nominees who held three-fifths of the shares. Under the Gauge Act, once a line had been used by passengers the gauge could not be changed except by authority of Parliament.

A difficulty still faced the LSWR: the gauge of the Taw Vale Extension, comprising 31 of the 37 miles between Cowley Bridge and Barnstaple, remained to be fixed by the Railway Commissioners. These gentlemen pointed out that they would not have their hands forced when the LSWR opened the original Taw Vale (Fremington–Barnstaple, for goods only) as a narrow line to the north and also narrowed the unopened Crediton line to the south. In impudent reply the Commissioners were told that narrow trains would start from Crediton to a temporary station at Cowley Bridge on 15 February 1848.

But on 8 February the Commissioners said the Taw Vale Extension should be broad. Reference was made to the foul means adopted to secure its Act, and to 'transactions of an extraordinary kind' affecting the Exeter & Crediton, especially the 'illegitimate application of the capital of the South Western Company, more particularly objectionable in a case where the funds have been misapplied for the purpose of controlling the Directors and Shareholders of an independent Company'.

Upon this the LSWR petulantly cancelled the Crediton opening, the narrow tracks being left to rust as the broad ones before, and stopped the construction of the Taw Vale Extension. This had just started when the Commissioners made their pronouncement. The threat of continued isolation for North Devon resulted in trade depression and large-scale emigration.

Barnstaple raged and the Press poured forth invective, but the stalemate continued until the LSWR reached one of its periodic decisions not to trouble to invade the West. Then the Exeter & Crediton was, after all, leased to the Bristol & Exeter, reconverted to broad and opened on 12 May 1851. The same year the Taw Vale Extension took the name of its dead rival, the North Devon Railway. The first sod of the resumed work was cut in February 1852 – at Copplestone, highest point on the line and suggested junction for various proposed railways to West Devon. In April rails began to arrive at Fremington: light goods traffic, horse-drawn until 2 November 1855, had been worked from here to Barnstaple since August 1848.

For most of the way the North Devon line follows the Taw as it gradually drops to join the estuary at Barnstaple, engineering works being confined to a few bridges. The contractor, Thomas Brassey, worked quickly, but aroused hope in Barnstaple needlessly early. For weeks, church bazaars and other events were announced for 'whatever day shall be fixed for the opening of the railway'. 12 July 1854 was the official date, but the Board of Trade required more signals and the regular service of four week-day and two Sunday trains did not start till 1 August. Rumours suggested the line was dangerous: only 80 passengers went to Exeter the first day. But soon Barnstaple's monotony was 'relieved by the passing and repassing of the new omnibuses and other conveyances which now ply between the hotels, the station, and different parts of the borough'. On 2 November 1855 passenger trains were extended to Bideford, the Bideford Extension Railway – virtually part of the North Devon – consisting of the six and a half miles from Fremington. The gauge between Barnstaple and Fremington had, of course, been broadened.

The Bristol & Exeter continued its lease of the Exeter & Crediton, and Brassey himself leased the North Devon and the Bideford Extension lines, the last two being worked as one. The LSWR took over the lease of the Exeter & Crediton from 1 February 1862, the Bideford Extension from 1 August 1862 and the North Devon from 1 January 1863. (The two latter were amalgamated with the LSWR on 1 January 1865.) The gauge was mixed in readiness for the extension of narrow trains to Bideford on 1 March 1863: they had been running as far as Crediton for some months. Most of the North Devon's varied rolling-stock was then sold, the coaches to the South Devon, but several locomotives were kept to work the daily broad-gauge goods which visited Bideford till 1877. The Bristol & Exeter and GWR retained goods rights at Crediton till 1903, the daily train being broad until 1892.

At first the track was single, except between Cowley Bridge and Crediton, and crossing trains involved setting one of them back into a siding. Crediton to Coleford Junction (a mile north of Yeoford station) was doubled as part of the improvement of the LSWR route to Plymouth in the late 1870s, and early this century Coleford Junction–Copplestone and Umberleigh–Barnstaple Junction were doubled. Powers for doubling the rest were obtained in 1906, but in 1911 the GWR and LSWR decided to cease spending on their rival routes to Barnstaple. This agreement caused more harm than the original delay in building the Vicar of Bray Railway (nicknamed after the sixteenth-century clergyman, twice Papist, twice Protestant). The bridges for a double track were there for the using decade by decade, but whereas the GWR improved its Barnstaple line in the 1930s, the Southern failed even to lengthen the loops at stations like South Molton Road (now King's Nympton). Crossing full-length trains at such places involved shunting back the rear coaches of one into a siding, points being locked by hand. Nor did the single track merely slow the journey: trains remained infrequent and, because of the expense of staffing the signal boxes at all the crossing stations, the service stopped early in the evening.

The last paragraph follows the wording of the 1958 edition except that it has been put into the past tense. True, even since nationalization there was a period when doubling was desperately needed to help link North Devon more firmly with the rest of the world. British Railways twice rejected appeals that it should be done, but on the second occasion, in 1957, the Southern Region did promise to lengthen loops and introduce centralized traffic control. But that had not happened before the Western Region took over in 1962, and thereafter the question was not what development would be undertaken but how much of the withered arm would be spared the axe. Those sections doubled between the wars have now all been singled, and most of the crossing loops have been abolished for good measure, leaving a few long single-line sections. Trains no longer have to set their rear coaches back into a siding at the remaining crossing places since most services consist of a three-car multiple unit and only one through service from Paddington reached Barnstaple even on summer Saturdays in 1973.

Barnstaple at least remains on the railway. In 1965 it looked as though the whole system north of Exeter would be closed as three bridges close to the junction with the Western's main line at Cowley Bridge needed replacing; but they were rebuilt, this being the first piece of double track to be singled. So Barnstaple still has its nine daily trains, though irritatingly all intermediate stations remain open and the pace of some services is distinctly slow. All the rest of North Devon's system has been closed to passengers.

As has already been implied, the bad service from Exeter to Barnstaple resulted in few links – trade, social or cultural – being formed between North and South Devon during the last century or this; the north remains a world strangely apart. The majority of passengers on Exeter–Barnstaple trains have always come from beyond Exeter, commuting business with both Exeter and Barnstaple being slight, though for many years a one-coach train starting at South Molton Road called at all stations to Barnstaple carrying schoolchildren and workers. Crediton, the most important intermediate place, has little to do with the railway – the railway which last century allowed Exeter to capture the market and other trade. Like most Devon inland towns, Crediton has declined by a third of its 1851 population. Chulmleigh, by-passed, was even more unfortunate, its market closing when one was opened at Eggesford station. Even since nationalization Chulmleigh begged for a halt between stations to be added at the nearest point to itself. Eggesford, incidentally, was chosen for the first State Forest, established in 1919. Today, trimly massed conifers are a prominent feature of the valley, but forestry has never provided much traffic. A creamery at Lapford provided heavy traffic for decades but was due to close in 1973 as part of a general scheme to turn more of the West Country's milk into dairy products, in this case at a cheese factory opening in North Tawton.

Better communications with the outside world would have helped all North Devon, but Barnstaple has been partly consoled, the isolation enabling it to retain administrative and other services which might have gone to Exeter. Barnstaple's population has doubled since the railway's arrival, the most spectacular increase being in the first quarter of this century, when the train service was at its best, at least in relation to the requirements of the age. Market Day (Friday) business at this time involved lengthening trains on all five routes leading to the town. Little of this local rail passenger movement survives, but it was the railway which allowed Barnstaple to become capital of North Devon, relegating Bideford to a back seat. The railway established the trend still followed by road transport: though the total amount of travelling has increased, the relative importance of routes is substantially the same as before 1914, though Ilfracombe's importance has grown.

Even pre-1914, full-length holiday trains were sometimes run to the resort, but normally Barnstaple Junction gave priority to the Bideford and Torrington sections of expresses. The extension from Bideford to Torrington was brought into use on 18 July 1872; owing to disputes elsewhere the LSWR had sought to withdraw the scheme, but Parliament insisted on it. It is not, of course, hard to see why the Bideford line lost patronage: at both Barnstaple and Bideford the stations were on the wrong side of the estuary, and the train service was poor, winter and summer. For decades, most Barnstaple–Torrington trains consisted of a single coach, albeit that may have been a through one from Waterloo. Closure to passengers was on 4 October 1965. Bideford once handled heavy general freight which disappeared when all freight was concentrated at Barnstaple, and the line also carried substantial quantities of three special traffics: coal, clay and milk. 50,000 tons of coal annually used to be imported for much of

North Devon's use (including locomotive coal) at Fremington Quay, modernized by the Southern Railway between the wars when it was the busiest port in Devon, second only to Plymouth. When the Western Region took over the Southern lines west of Exeter, the traffic was diverted *via* the Severn Tunnel. Other coal was brought from 'up country' into Yelland power station near Fremington, but though this remains rail-connected, for many years most has come *via* sea. Clay comes from two works on the North Devon & Cornwall Junction line between Torrington and Halwill: at Petrockstowe and Marland, now the terminus. Until the closure of Fremington Quay on 30 March 1970, much of the clay was exported there; heavy traffic still goes to Fowey for export and *via* rail all the way to the Potteries and Scotland. Clay tonnage has probably averaged 30,000 this century. Milk, even at the time of writing, comes from the West Country's largest creamery at Torrington; in 1957 nearly 2,000,000 gallons went to London. But the emphasis has increasingly been on manufacturing the milk into dairy products and it was anticipated that no liquid milk would be sent to London to help out with the 1973 peak demand.

Ilfracombe was the subject of numerous unsuccessful schemes, including rival LSWR and Devon & Somerset Bills in 1864. An eventual agreement for a joint line of mixed gauge broke down when the Devon & Somerset could not pay its share. At length a new approach was made by the LSWR, a subsidiary Barnstaple & Ilfracombe Railway receiving Royal Assent on 4 July 1870. Now the price had to be paid for the railway's late entry into North Devon. So depressed had the economy become and so many labourers had emigrated that not enough navvies could be found and only on 20 July 1874 was the line of violent gradients opened. Down trains climbed three miles at 1 in 40 and then dropped at 1 in 36 for two miles round sharp curves to the terminus perched spectacularly on a cliff edge above the town. Engineering works were heavy, with tunnelling and the S-shaped bridge over the Taw at Barnstaple as well as almost continuous embankments and cuttings, except in the section beside the Taw estuary. The view just short of Ilfracombe, as the nervous passenger prayed the brakes would grip, was magnificent.

A special class of passenger locomotives, 0-6-0s with outside cylinders, was designed by Beattie senior, locomotive engineer of the LSWR, but he died in the opening year and his son rejected it: the first of the famous 'Ilfracombe Goods' engines were ready, and they were allowed to work all traffic for a long period. 0-6-0s with inside frames and cylinders, open splashers and a flush-top boiler, they were of a light-goods class Beyer Peacock & Company supplied to overseas railways with heavy gradients, notably in Sweden. During its first years the Ilfracombe line had lightweight rails, and only selected rolling stock could be used.

The year before the opening the Devon & Somerset had bestirred itself and inaugurated a coach service from its Barnstaple terminus (Victoria Road) to Ilfracombe. This continued in hot opposition to the LSWR until 1 June 1887, the opening date of the mile-long loop from the terminus round the outskirts of the town to Barnstaple Junction. At first it was officiously declared:

> The Ilfracombe Through Carriages must be reserved for Passengers going to the Ilfracombe Branch, and Passengers for other Stations on the L.S.W. must not be allowed to use them ... but must ... proceed across the town to the L.S.W. Quay or Junction Stations as hitherto.

Barnstaple Junction – or just Barnstaple till the opening of the Quay station on the Ilfracombe line – had its down island platforms added in May 1874 in readiness for that event. Barnstaple Quay was replaced by Town on a larger adjoining site in time for the Lynton line's opening in 1892. Barnstaple (but not the bridge over the Taw, which remained single) to Ilfracombe was doubled in three stages, in 1889–91. The picture was then complete except for the opening on 1 July 1905 of the short Barnstaple eastern spur, which completed a triangular junction outside the GWR's station (later Victoria Road) and permitted Taunton–Ilfracombe trains to by-pass it. The spur was closed in 1939, all trains again having to reverse at Victoria Road until that was closed to passengers on 13 June 1960. The rebuilt eastern spur then formed part of one continuous section from Swimbridge to Barnstaple until the closure of the Taunton–Barnstaple route in 1966: the track from the Barnstaple Junction direction into Victoria Road goods was relegated to siding status until Victoria Road depot itself closed from 5 March 1970.

For a long time the GWR played an important part in Ilfracombe's development and maintained a town office there. A few summer Saturday through trains continued to reach Ilfracombe *via* the former GW route into the early 1960s, but most passengers were routed *via* Exeter, involving an extra 22 miles – a sorry state of affairs. One thing alone saved Ilfracombe between the wars – the old-established steam-packet service which brought Welsh holidaymakers across the Bristol Channel in hundreds, especially before about 1930, when Welsh resorts began to compete. Colourful are the stories of Welshmen on holiday rampage; street concerts to collect the return fare were common. The sea crossing was frequently rough and the GW gave undue preference to Ilfracombe–South Wales trains, hoping to tempt those seasick on the outward journey. Bad rail services meant that in the 1950s and 1960s Ilfracombe received far more passengers by motor coach than did other resorts of similar size, while growth in the holiday trade was poor and the season shorter than South Devon's. Even so, it was often block-to-block working with double-headed and banked trains over the gradients between Barnstaple and Ilfracombe, until the route was first singled and then closed, from 5 October 1970. Though rail traffic had been dwindling – nay positively discouraged – the final withdrawal hurt many hotels and boarding houses and led to unsuccessful attempts to get the line reopened on a private basis. Goods traffic – shades of the 'Ilfracombe Goods' – ceased from 7 September 1964 when Rolles Quay at Barnstaple also lost its rail connection.

So today Barnstaple is the end of the line apart from the continuation through Torrington to Marland used only by clay traffic. Miraculously, the nine passenger and two freight trains that run on most days have been carrying increasing traffic during the past few years, and with a new industrial estate opening at Barnstaple the future could be assured – apart from another administrative change which withdraws all top management from North Devon. With nobody to fight for traffic such as the steel used in a revitalized Appledore shipyard, the withered arm may yet be pruned out of existence.

TO OKEHAMPTON, LYDFORD AND NORTH CORNWALL

An account of the first schemes for a narrow central route through Okehampton to the far West may be left to the Cornish chapter. Here we begin in 1862, when the Okehampton Railway was incorporated to build a narrow-gauge line from Coleford on the North Devon Railway to Okehampton, and when the LSWR signed an agreement with the GWR and South Devon Railways. This included a definition of the companies' territories and indicated that the LSWR was not to go, or have any interest, beyond Okehampton. In return the Andover–Redbridge line was transferred to the LSWR (see Vol. II) Next year the Okehampton Railway obtained powers for an extension to Lydford (then spelt Lidford) on the broad-gauge Launceston & South Devon Railway, to be given a third rail to allow narrow trains to reach Plymouth. Sections permitting the extension to be worked by the LSWR were withdrawn from the Bill when the South Devon's chairman pointed out that such an arrangement would be contrary to the new non-aggression pact.

Though the LSWR's sincerity in signing the agreement is questionable, for a time the Okehampton Railway at least gave the impression of independence. In 1865 it changed its name to the Devon & Cornwall, adding powers for extension to Bude and Torrington to those it already had for Lydford. Another railway to change its name this year was the Launceston, Bodmin & Wadebridge Junction, incorporated in 1864, which became the Cornwall Central and extended its powers to include a line to Truro. Between them the two companies promised a narrow-gauge route linking Waterloo not only with the LSWR's remote possession, the Bodmin & Wadebridge, but also with the extreme west of Cornwall, the gauge of the West Cornwall Railway, from Truro, being mixed. This, plus the temptation of Plymouth, was understandably too much for the LSWR to resist, whether or not it had secretly encouraged the expansion of the two companies.

The result was an appeal to the GWR and South Devon to revise the 1862 agreement. Numerous meetings were held and this time there was also a fourth party – the Bristol & Exeter, whose inclusion proved the stumbling block. A 'Quadruple Agreement' provided for various traffic arrangements between the companies, and conceded the LSWR the right to go to Plymouth on the understanding that it was to be at a disadvantage concerning terms. The new arrangement was to be cemented by a Parliamentary Bill in 1867, but the LSWR withdrew on the pretext of a dispute with the Bristol & Exeter over North Devon traffic, and then used this dispute to repudiate the 1862 agreement. The Bristol & Exeter, of course, had not even been a party to it! Later, efforts were made to ratify a modified Quadruple Agreement, favoured by the LSWR shareholders, despite the advice of their general manager, Archibald Scott, to continue aggression. But, fear of monopolies being rife, Parliament was in the mood to prefer competition to co-operation, and would not look at the Bill. The broad gauge therefore had no protection against the impending invasion of its most cherished reserves. In this chapter the LSWR is inevitably the villain; but when we arrive at Plymouth it will be seen that through lack of enterprise the broad-gauge companies were at least partially responsible for their fate.

The Devon & Cornwall was officially treated as a subsidiary from 1866 and absorbed by the LSWR in 1872. The first section, from Coleford Junction to North Tawton, had been opened as early as 1 November 1865, but owing to financial stringency progress was slow. North Tawton to Okehampton Road was ready on 8 January 1867, the last three miles to Okehampton on 3 October 1871 (Okehampton Road was named Belstone Corner, but in 1872 became Sampford Courtenay), and the extension to Lydford on 12 October 1874.

The maxim of the early main-line builders had been to go over or through natural obstacles; here the policy was to dodge them. The first section from Coleford Junction is simple enough, and includes four straight miles beyond Bow; but soon frequent curving is needed as the track climbs (often at 1 in 77) and clings to the periphery of Dartmoor. Okehampton station looks down on the town from the moorside. Two miles later, just beyond Meldon Viaduct, the maximum height of 950 ft is reached. Then followed a gracious series of curves across open moor – best seen on a sparkling spring morning – as the descent, mostly at 1 in 80, began to Lydford and beyond.

At first the track was single, but doubling proceeded quickly after Lydford had been reached, and was completed – including

Meldon Viaduct – by the end of 1879. (It was not until 1890 that the narrow gauge had its independent double line from Lydford to Plymouth.) Meldon Viaduct crosses the valley of the West Okement, is built on a curve of 30 chains, and has six girder spans supported by metal lattice piers, the tallest being 120 ft. Its appearance is slight, and aesthetically something looking readier to withstand the elements, preferably built of local materials, would be more in harmony with the moor. Meldon Junction, of which more in a moment, was to the west; to the east is the railway's own Meldon limestone quarry, one of the largest in the country with an annual production of over 300,000 tons, much used for ballast and conveyed in hopper waggons by regular daily trains.

Apart from ballast, traffic between Coleford Junction and Lydford was once mainly of a through nature but after the 1963 boundary adjustments the Western Region relegated the route. It was closed completely between Okehampton and Bere Alston from 6 May 1968, whereafter a shuttle service operated between Exeter and Okehampton. This too was closed, from 5 June 1972. The route had already been singled and today's freight and mineral branch begins at Crediton, running beside but independently from the Barnstaple line as far as the former Yeoford Junction. Okehampton had never been railway minded, although the grim little town went gay enough when the first train arrived. People eating the ceremonial cream tea in the main street regarded overdue progress as imminent: in fact the railway's first accomplishment was to transfer much trade to Exeter. Okehampton was disillusioned, but as a junction town it did stretch its influence over a wider surrounding area and ticked over until new opportunities came when motor traffic reawakened the main road to Cornwall. Neighbouring places could not even remain static. Take North Tawton. Until 1865 it was almost Okehampton's size, with a woollen and a flour mill. Both mills closed upon the railway's arrival; shops lost trade as the unemployed mill workers drifted away and big shops elsewhere attracted local custom. Agriculture itself was in the doldrums, especially after 1875, and the position was complicated by the transition from crops to dairying, at least partly encouraged by the railway's ability to take milk quickly and cheaply to the centres of population. By 1889 for the first time Devon's pastureland exceeded that under cultivation. In addition the railway stimulated the spread of new farming ideas and made the use of artificial manures easier.

Much of the Devon–Cornwall border country's transport history is concerned with manure. When the upper Tamar was improved, it was known as the Tamar Manure Navigation, while the Bude Canal, the most remarkable engineering work of its day, using boats which fitted on rails to mount the inclined planes between two levels of water, carried principally sea sand for manure. With cheap limy sand available, a strip of land a few miles deep on either side of the canal became noticeably richer than the rest of the area; similarly, when artificial manures grew fairly common the land round railway stations increased its value. Artificial manures were, of course, on the market well before the railway approached the Atlantic; it was, indeed, partly because of the demand for them in this poor country that the railway was encouraged.

The 1865 powers for Bude lapsed and for the time being it was decided to go only as far as Holsworthy, the single line from Meldon Junction being authorized in 1873 and opened on 20 January 1879. The area's population was already depleted, but in its early years the booking-office must have issued a startling number of single tickets. Holsworthy acted as railhead and start for the LSWR's smart coach service until 10 August 1898, when Bude, a declining port but growing resort, at last welcomed its first train. The line was built by the LSWR; a local company which obtained its Act in 1893 failed to make progress. Holsworthy Viaduct, of nine 50-ft arches, was the first work of such size to be built in concrete, and the new section's chief engineering feat. The terminus was deliberately placed on Bude's outskirts to placate Stratton, but a branch was built down to the canal basin. The railway replaced the canal for much of the sand carriage, and a horse-drawn flanged rail tramway was built from the beach to a tip above the standard-gauge siding. The tramway was later rebuilt of ordinary rails and may still be seen, though for a generation all the sand has been taken by lorry direct to the farms.

Bude is isolated, but would surely be twice its present size had the railway arrived earlier and given a better—particularly a quicker – service. The branch is steeply graded, though the district is more plateau-like than hilly: the LSWR did not trouble with much embankment and cutting making. Apart from the slow gradient climbing, long delays occurred at Exeter, Okehampton and Halwill while the various sections of trains such as the *Atlantic Coast Express* were divided: that was in steam days, when Halwill, roughly half-way between Meldon Junction and Holsworthy, was an oasis of activity (shunting always seemed to be in progress) in a hard, poor and sparsely-populated country where life almost stands still. The station, and the small village named after it, remains the economic and social centre for miles around, but first all freight was withdrawn, leaving only occasional diesel units, and then came total closure of the whole system west of Okehampton, including Holwill–Launceston–Wadebridge, on 3 October 1966.

The 1864 powers for the Launceston, Bodmin & Wadebridge Junction had lapsed many years before the LSWR at last felt ready to bridge the gap between its main system and the Bodmin & Wadebridge. Not until 18 August 1882 did the subsidiary North Cornwall Railway obtain its Act, and then construction across the bleak countryside was painfully slow, due to money shortage. Halwill to Launceston was opened on 21 July 1886. Tresmeer was reached after another six years, on 28 July 1892. Tresmeer–Camelford followed on 14 August 1893, Camelford–Delabole on 18 October 1893, and Delabole–Wadebridge on 1 June 1895. Thus was North Cornwall given the railway it had despaired of seeing. The advent of trains – the line's service was more frequent and often faster in the early days than now – must have meant revolution to those featureless inland villages which are the reverse of what most people picture as 'Cornwall'. Camelford awoke from a half-century's slumber; for Delabole the railway brought even more – greatly extended business for the slate quarry, the largest in the world: so grateful were its owners when the line at last approached that they gave land free for three-quarters of a mile of track. From the train you could peer down into the gigantic hole, and see the men at work looking like ants and the trucks being pulled up the narrow-gauge tramroads. Such slate traffic as the railways retain is now handled elsewhere, but once the North Cornwall line yielded useful freight, such as 750 containers of fresh meat and 625 trucks of cattle at Launceston in 1957. Local passenger business was always thin, but the line had its great moments – summer Saturdays when the crossing loops at all the intermediate stations proved justified. 'West Country' class locomotives toiled up the long miles, mainly at 1 in 73 to 1 in 78, on either side of the summit section between Otterham and Camelford, pulling trains crowded with visitors going to Padstow and a large number of small resorts beyond the railway system.

BRANCH AND MINOR LINES

The LYNTON & BARNSTAPLE, the West Country's only genuine narrow-gauge (1 ft 11½ in.) passenger-carrying railway, was sponsored by local people after earlier schemes to link the twin villages of Lynton and Lynmouth with the outside world failed to mature. Hitherto the GWR had refused to show interest, but no sooner had the small Company been formed than Paddington produced a rival plan. Nothing came of it, but the threat gave Lynton & Barnstaple proprietors the first of their many worries. The next came when the contractor claimed an extra £27,000, apparently merely because he had not expected to have to excavate any rock when building the track across Exmoor's foothills. Construction worked out at £5,000 instead of the estimated £2,500 a mile. Lengthy litigation decided against the contractor, but as that bankrupted him the Company had to pay the expenses as well as complete the line, opened on 16 May 1898, a year later than promised. From the start traffic fell hopelessly short of expectations, the position of the Lynton terminus, several hundred feet above the village, discouraging business even before the coming of motors.

In April 1903, Sir George Newnes, leading light among the shareholders, shocked the neighbourhood by announcing he was buying two vehicles to ply between Blackmoor station and Ilfracombe. Shortly after the service began the local Press announced: 'One of these cars was travelling at a little over 8 m.p.h. on a bye-road when the police interfered, a prosecution was instituted and a heavy fine inflicted.' Upon that the vehicles were sold to the GWR for its pioneer Helston–Lizard service. Meanwhile the Lynton & Barnstaple was costing too much in repairs as well as carrying too little traffic. Trains always rolled badly, whether or not the track was in repair. It was officially declared: 'Not the slightest apprehension of danger need be felt.' Not until 1913 was the first dividend paid on ordinary shares: ½ per cent per annum was the rule until 1921, when distribution ceased. In 1922 a loss was made, and though the Company was unaffected by the 1921 Railway Act it sold itself to the Southern Railway. The total outstanding capital was £127,000; the purchase price £30,000.

The track, rolling stock and train service were immediately improved. Normally five trains ran daily, many of them mixed, taking 100 minutes for the 19 miles, including time performed in shunting at the five crossing stations, all fully signalled. Double-headed trains were once needed to take shoppers home from Barnstaple on Market Days, but when quicker buses arrived passenger traffic became negligible except during the short holiday season. The last train, double-headed and carrying 300 people, ran on 29 September 1935. At the auction held at the railway's extensive Pilton headquarters at Barnstaple, five well-kept engines fetched a mere £236 together. The remaining engine, *Lew*, began hauling the demolition trains. Lynton became about the farthest place in England from a station; coal – the one traffic that had fully reached the railway's expectation – shot up in price. Interest in narrow-gauge lines has increased so much that if the Lynton & Barnstaple still existed, its future would be secure, though it would probably run only during the summer. Indeed, enthusiasts steadily trickle to Chelfham village and gaze up at the viaduct trying to evoke a narrow-gauge locomotive. The viaduct, of eight arches 24 ft wide, was the chief engineering work and has been allowed to stand as a reminder of the local people's enterprise.

The BIDEFORD, WESTWARD HO! & APPLEDORE Railway, another highly individual line, had an even less happy and much shorter career. It was opened from the quayside station at Bideford to Northam on 20 May 1901, and extended to Appledore, a small port and shipbuilding town, on 1 May 1908. The extension was built as a Light Railway and the original section was also worked under the Order, enabling gates to be dispensed with at five level crossings. There were three locomotives: *Grenville*, *Kingsley* and *Torridge*, 2-4-0 tanks with their motion enclosed and provided with cowcatchers as the highway was followed on the first section out of Bideford; and six American-style polished teak passenger-cars, elaborately decorated inside and out. Normally the seven to fourteen daily trains made three regular and eight conditional stops (including Lovers Lane) as they rambled across the district, then little developed but today noted for its caravan camps. Traffic was occasionally very heavy, but on balance was insufficient and the Company lost steadily and could not even manage the ½ per cent. dividend of the Lynton & Barnstaple. Closure would anyway have become inevitable, but the Government commandeered the track, trains running for the last time on 27 March 1917. Though of standard gauge the railway was an isolated concern. After the closure a temporary track was laid across Bideford Bridge to allow the locomotives to reach the LSWR.

The NORTH DEVON & CORNWALL JUNCTION Light Railway, between Torrington and Halwill, was opened on 27 July 1925 and was the last line of any length to be built in the West. The first six miles from Torrington replaced a 3-ft gauge tramway, built to carry clay from the North Devon Clay Company's Marland works. The rail entrance to these works is near Dunsbear Halt: an extensive system of 3-ft gauge sidings, worked by diesel locomotives, is retained inside. Shortly beyond Petrockstow is access to another clay works, again with its own railway system, above and below ground. The clay has always provided the standard-gauge line with its chief traffic. The largest place on the railway is Hatherleigh, 13 miles from Torrington. A market town in the centre of a barren countryside, it is badly decayed – and was when the line was under construction and hundreds of navvies arrived on the scene. Among the last of the vanishing railway-building race, the men quickly won the respect of local people, but unemployed labourers sent to work alongside them deliberately created trouble in the hope of being sacked and returned to the dole. Hatherleigh saw the liveliest night in its history when a vicious attack was made on the police. After that the place forgot the railway's passenger trains; the station was two miles away and the train journey to the nearest larger town, Okehampton, involved twenty miles compared with seven by road. But there was fair cattle traffic until British Railways reduced their livestock handling arrangements in 1963. Passenger services were belatedly withdrawn on 1 March 1965, since when only the Marland–Torrington section has been retained, for clay traffic.

At the end of the 1870s extensive peat works were opened between Great Links Tor and Kitty Tor on the north-western slopes of Dartmoor. The Bridestowe or RATTLEBROOK PEAT railway, built in 1879, began from sidings laid beside the LSWR slightly north of Bridestowe station, about 750 feet above sea-level, and climbed in a zigzag course to beyond the 1,800-ft contour. As the crow flies the distance is 2½ miles, but the railway took nearly 6 miles, including a reversal point. It was worked by a steam locomotive, but no details are known. Like so many moorland industrial ventures, the Dartmoor Compressed Peat Company quickly ran into financial troubles, and the railway had only a short continuous career, though it was used spasmodically into this century. Then, during the 1914–18 war, chemical works were established on the moor, making use of the peat. A petrol-driven truck took out building materials and machinery: sometimes it also pulled a standard goods truck. The big problem was to dry the peat sufficiently, and the truck took men out on night as well as day shifts to keep up heat in the retort houses. But this scheme proved no more successful than the first, and after the war the truck brought back what could be salvaged from the works, and finally was used to dismantle the railway in 1931–2. Since then the railway's course has occasionally been used by lorries to bring down peat, last loaded at Bridestowe station for Guernsey in about 1955.

I have been unable to discover details of another Dartmoor peat line, the isolated ZEAL TOR tramroad, which ran 3½ miles from Shipley (north of South Brent). Narrow gauge, with rails bolted to timber, it was working at the turn of the century.

CHAPTER VI

Plymouth and District

THE PLYMOUTH & DARTMOOR RAILWAY

Sir Thomas Tyrwhitt, successful politician and diplomat, spent much of his later life and all his money trying to cultivate and populate Dartmoor. After twenty fruitless years, he persuaded the Admiralty that the site of Prince's Town, named after the Prince of Wales (George IV), was suitable for a 'depot' for French prisoners. Dartmoor Prison was finished in 1809; 5,000 prisoners once lived there, but after the 1815 peace the buildings were deserted. Only then, in 1818, did Tyrwhitt think of a railway, from Crabtree on the Laira (Plym estuary) to Princetown. He wrote:

> To reclaim, and clothe with grain and grasses a spacious tract of land now lying barren, desolate, and neglected; to fill this unoccupied region with an industrious and hardy population; to create a profitable interchange of useful commodities between an improvable and extensive line of backcountry and a commercial seaport of the first capabilities, both natural and artificial; to provide employment for the poor of several parishes; and to alleviate the pressure of parochial burdens by a method at once simply ingenious, and comparatively inexpensive, form altogether such a stimulus to adventure, and such a scope for exertion, specially to a wealthy company, as must dilate the benevolent heart of the patriot, whilst it emboldens the capitalist gladly to lend his assistance in carrying the plan into execution.

Granite from King Tor, near Princetown, was to be the chief traffic, but Dartmoor peat could have many uses. Lime, sea sand for manure, coal and timber were expected to fill the upward trains, together with tea and sugar which 'are become absolute necessaries in the present day'. The prospectus, published with an account of a public meeting held on 29 March 1819, estimated an annual revenue of £2,930.

The Plymouth & Dartmoor had three Acts. The first (2 July 1819) authorized the Crabtree–Princetown line with £27,783 capital and £5,000 additional borrowing powers. Canal terms, such as 'wharf' for 'station', were used. The second (8 July 1820) authorized an extension to Sutton Pool with a branch to Cattewater, together with another £7,000 and powers to borrow from the Commissioners of Public Works. The third (2 July 1821) explained that the line between Crabtree and Jump (Roborough) 'could be made with a regular Ascent and upon a much better Construction.' Deviations to follow the contour were allowed – and a 620-yd tunnel.

26 September 1823 was opening day for the 23 miles of single 4 ft 6 in. track, said to cost £66,000, from Sutton Pool to King Tor, a distance as the crow flies of roughly 13 miles. About 1,000 people were invited by Tyrwhitt to breakfast at the Roborough Down wharf, and then a procession of wagons, carrying granite and passengers, set out in the rain, trundled through the tunnel cut through solid rock, and was given a 'rousing welcome from the assembled multitudes' at Plymouth.

Financial trouble started at once. The Company lacked money to finish the line at the Princetown end, and it was completed by Messrs John and William Johnson. They were working the quarries and needed the railway to carry granite for their Plymouth contracts. The Press speaks of their 'unparalleled liberality', but naturally they acquired a mortgage in the Company to offset the cost incurred. As the Company could not pay interest on this mortgage, the Johnsons paid no tolls on their granite traffic. Other traffic was scarce – we hear no more of Tyrwhitt or his promises, though he was the largest shareholder – and as time went by the need to reach a compromise with the Johnsons was stressed. But 'No progress' is a frequent entry in the minutes, until they ceased to be kept for several years, after the dramatic entry of 22 May 1847: 'Broken off negotiations owing to the unexpected unreasonable demands' of the brothers.

By an Act of 19 June 1865, William Johnson (John was dead) received £75,000 in 5 per cent preference shares in exchange for the mortgage, and the Plymouth & Dartmoor became its own master again. A few years later it was using its granite traffic revenue to relay much of the line. But the route above Yelverton was soon taken over by the standard gauge Princetown Railway. Only occasional traffic then used the lower part of the line, till about 1900. Rails, except between the Rising Sun and Plymouth, which section was incorporated in the Lee Moor tramway, were removed in 1916. In some places the route is now difficult to follow, but walking it with the aid of a large-scale map can be rewarding. The two portals of the tunnel are intact, but it is not possible to go through.

The Plymouth and Dartmoor could not afford to add to its system itself, but it (or three of its directors as individuals) should have built the branch to Cann Quarry, required by Lord Morley of Saltram, an opportunist, as payment for the withdrawal of his opposition to the Company's 1821 Act. The Plymouth Bank failure bankrupted two of these directors, and the third, John Pridham, on whom fell the whole responsibility, was Morley's solicitor. His Lordship instituted proceedings against him – but eventually built the branch himself. This episode is typical of many in the story of the 4 ft 6 in. gauge; details of the branches are summarized in Appendix II.

THE MAIN LINE ARRIVES

In 1821 Plymouth's population was 21,000. Twenty years later it was 36,500, but the town had expanded little in area. A third of the people lived in single-room homes, most throwing their sewage into the street. The Improvement Commissioners were impotent. During the early 1840s the situation went from bad to worse. There was a gay cultural life, true, but even the guide books had to admit something of the town's drawbacks. Undoubtedly the late arrival of the main line delayed badly-needed reform. Meanwhile, Devonport, which had gained independence in 1837, was bigger if slightly less squalid than Plymouth, and between these two boroughs Stonehouse, the third of what are termed locally the Three Towns, had already crowded in 10,000 people.

Page 113 Sea wall—1

Plate (25) The sea wall at Dawlish and Teignmouth has probably been the most painted, sketched and photographed section of railway line in Britain. The first of this selection from the hundreds of available pictures is a photograph by permission of A. J. West of Dawlish, from the original water-colour by W. J. Dawson in 1848 entitled 'The line of the South Devon Railway at Dawlish'. Note the pumping station and its tower.

Page 114 Sea wall—2

Plate (26) Landslip at Breeches Rock in December 1852 (from a contemporary lithograph). Note the third-class carriage.

Plate (27) A breach of the sea wall outside the Teignmouth end of Parson's Tunnel in February 1855.

Page 115 Sea wall—3

Plate (28) An outing of John Heathcoat's Tiverton lace factory workers to Teignmouth on 10 August 1854. Note the signal.

Plate (29) The 'Cornish Riviera Express' at the same spot the first day it was diesel-hauled, 19 June 1958.

Page 116 Sea wall—4

Plates (30 and 31) Two snaps at the century's turn.

Plate (32) A 1930 West-to-North express.

The South Devon Railway's 1844 Act had envisaged a terminus at Eldad, but soon it was decided to have two termini, one for Plymouth (north of Union Street) and one at Devonport New Passage, where it was planned to transfer traffic by ferry to the Cornwall Railway. When the Cornwall dropped its ferry ideas, an inland Devonport station was chosen. Powers for the line leading to this were later transferred to the Cornwall Company. Finally the Board of Trade persuaded the South Devon to abandon a proposed goods level-crossing over Union Street; as a bridge capable of taking locomotives was to be built across it, the passenger station, too, was moved south to Millbay. The line from Laira to Millbay was opened on 2 April 1849.

If Plymothians crowding round the gaily decorated but undistinguished wooden terminus had known what lay in store, their rejoicings would have faded. In a few weeks cholera struck, killing nearly 2,000. The coincidence of the railway opening and this outbreak inevitably had profound effect on the leading citizens. It took several years to recover equilibrium, but in 1852 the old Improvement Commissioners were, to quote the Press, 'improved out of existence'. The year was a vital one for Plymouth as for the rest of South Devon. Newspapers were full of reports, correspondence and editorial comment about a new workhouse, a campaign to regulate shop hours, the need to reduce the number of public houses and beer shops, and about improvements in water supply and drainage. The most popular controversy concerned the merits of local *versus* Greenwich or 'railway' time: telegraphic communication with London was completed during the year, and had a deep psychological effect as well as greatly aiding the Admiralty, busily expanding its activities at Devonport. At Millbay work restarted on the 13-acre Great Western Docks, abandoned in the 1848 confusion. It was announced that eight steam packets would soon be diverted from Southampton to Plymouth.

1852 also brought renewed confidence to the Cornwall Railway – and renewed opposition to it from local people wanting Plymouth to remain the line's terminus, to prevent Falmouth competing as a port. By 1858 the railway from Cornwall Junction, half a mile north of Millbay, to the approach to the Royal Albert Bridge at Saltash was ready, including viaducts at Stonehouse Pool, Keyham and Weston Mill. Brunel's masterpiece carrying the railway high over the Tamar into Cornwall was not finished for another year. The river is 1,100 ft wide and 70 ft deep at high water, and the Admiralty required a high water head clearance of 100 ft. After excavations had revealed deep sand and mud, Brunel discarded a plan for a timber bridge with numerous spans. Even sinking piers for a four-span bridge was later deemed too difficult – but a single span too costly. So two main spans, each 455 ft, were designed. The bridge combines the suspension and the conventional type, the outward thrust of the arches on the abutments counteracting the inward drag of the chains. The two great wrought-iron tubes are arched upwards exactly to balance the fall of the suspension chains, while these chains, the tubes and the bridge platform itself are linked by vertical chains, on either side of the track. A double line had been planned at first, but the saving of £100,000 by having a single one was preferred. 730 yds long and with a total of seventeen approach spans as well

---

The map shows the situation as it was in 1958. Since then there has been major contraction. The former branch to Tavistock and Launceston terminates at the china-clay works; a cement factory at Plymstock is the end of the branch that used to run to Turnchapel, and the Yealmpton line has also been abandoned. The whole of the former Millbay complex south of the main line has been removed, though after the first edition of this work was published the old platforms at Millbay terminus were abolished to make room for additional carriage sidings, and a new diesel multiple unit depot was built between Millbay and Cornwall Junction. All locomotives are now stabled at Laira and carriage sidings built on the site of Laira marshalling yard for Southern Region trains are now the stabling point for Western rolling stock. Friary passenger station has been replaced by the city's main goods station, while North Road (now just called Plymouth) has a new skyscraper office block. The former South Western line between Devonport Junction and St Budeaux has been removed and the wartime connection at St Budeaux is now the junction: the South Western route no longer reaches Tavistock but terminates at Bere Alston where trains reverse for the stub to Gunnislake.

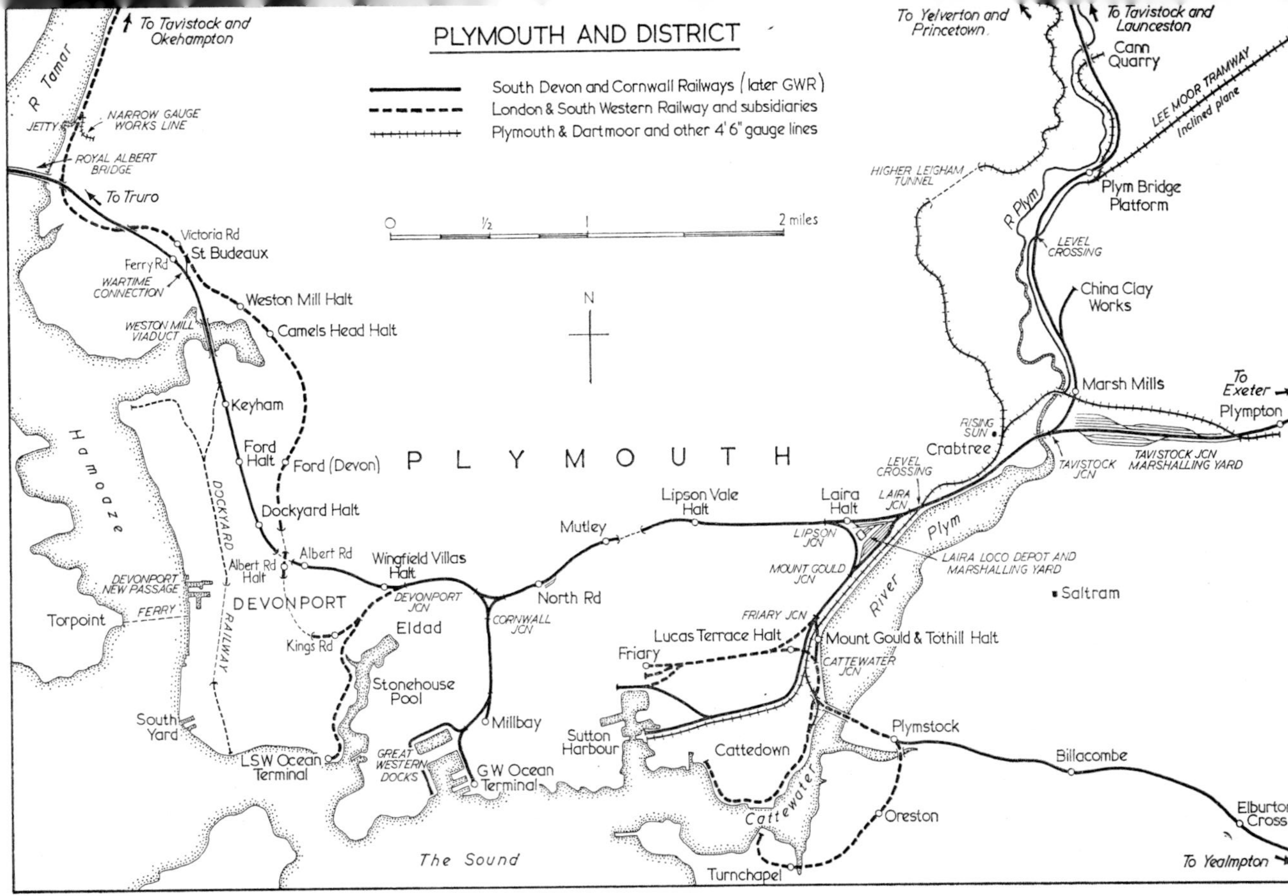
PLYMOUTH AND DISTRICT
South Devon and Cornwall Railways (later GWR)
London & South Western Railway and subsidiaries
Plymouth & Dartmoor and other 4'6" gauge lines
0
½
1
2 miles
N
PLYMOUTH
To Tavistock and Okehampton
R Tamar
JETTY
NARROW GAUGE WORKS LINE
ROYAL ALBERT BRIDGE
To Truro
Victoria Rd
St Budeaux
Ferry Rd
WARTIME CONNECTION
Weston Mill Halt
WESTON MILL VIADUCT
Camels Head Halt
Keyham
Ford Halt
Ford (Devon)
Dockyard Halt
Hamoaze
DOCKYARD
RAILWAY
Albert Rd
Albert Rd Halt
DEVONPORT
DEVONPORT NEW PASSAGE
FERRY
Torpoint
Wingfield Villas Halt
DEVONPORT JCN
Kings Rd
Eldad
Stonehouse Pool
CORNWALL JCN
North Rd
Mutley
Millbay
South Yard
LSW Ocean Terminal
GREAT WESTERN DOCKS
GW Ocean Terminal
The Sound
Lipson Vale Halt
Sutton Harbour
Friary
Lucas Terrace Halt
FRIARY JCN
MOUNT GOULD JCN
LIPSON JCN
Laira Halt
LAIRA JCN
LEVEL CROSSING
Crabtree
RISING SUN
Mount Gould & Tothill Halt
CATTEWATER JCN
Cattedown
Cattewater
Turnchapel
Oreston
Plymstock
Billacombe
Elburton Cross
To Yealmpton
River Plym
LAIRA LOCO DEPOT AND MARSHALLING YARD
Saltram
TAVISTOCK JCN
TAVISTOCK JCN MARSHALLING YARD
Plympton
To Exeter
Marsh Mills
China Clay Works
LEVEL CROSSING
R Plym
HIGHER LEIGHAM TUNNEL
To Yelverton and Princetown
To Tavistock and Launceston
Cann Quarry
LEE MOOR TRAMWAY
Inclined plane
Plym Bridge Platform

as the two main ones, the bridge cost £225,000. Prince Albert officially opened it on 2 May 1859. Public traffic began on 4 May. Brunel was then abroad: a dying man, the last he saw of the bridge was from an open truck drawn across specially for him.

1859 – 22 June – also saw the opening of the Tavistock branch, after years of controversy. As early as 1845 the South Devon had felt forced to propose its own line to avert narrow-gauge Central invasion; in 1848, similarly, the broad-gauge companies had subscribed to the Plymouth Great Western Dock Company in retaliation to an LSWR project linking a Plymouth branch *via* Tavistock with a Sutton Harbour scheme. Eventually the battle for Tavistock was fought between the South Devon and its ex-chairman, Thomas Gill, who resigned over the atmospheric fiasco. A scheme of his which had found support in North Devon and grandly became known as the Plymouth, Tavistock, Okehampton, North Devon & Exeter, was defeated in the 1853 session, when the South Devon & Tavistock passed the Commons to be rebuffed by the Lords. Next year Gill's party called itself the Plymouth, Tavistock & Devon Central, but on 24 July 1854 the South Devon's subsidiary won – with Lord Morley's help.

Engineering work on the 13 miles of single track from the junction with the main line near Marsh Mills was heavy, including Shaugh, Yelverton and Grenofen tunnels and six large timber viaducts (replaced 1893 to 1910) on stone piers, the longest being Walkham (367 yds long, 132 ft high), 'the most matured design of Brunel's timber viaducts'. The line was worked by the South Devon, with which it amalgamated in 1865.

Cornish people making long journeys began using the railway as soon as it reached Plymouth, travelling to it by the Torpoint ferry. Launceston, whose economy had been harnessed to the old road traffic, inevitably suffered. Plans for its own railway were numerous, but year after year nothing happened. The choice was between a line to Tavistock and Plymouth, and the Central to Exeter. Eventually the South Devon produced a subsidiary Launceston & South Devon Railway; to delay longer might have tempted the narrow gauge too much. The Bill was passed by the Lords on 30 May 1862. 'We well remember the trial of Queen Caroline . . . and the capture of Sebastapol, but on these occasions there was not the hundredth part felt by Launceston people as at present.' The scales had tipped in favour of the Bill by the Duke of Northumberland agreeing, after receiving a petition, to withdraw his opposition.

Official opening day for the 19 miles from Tavistock was 1 June 1865, a day so wet that even now local people talk of 'railway weather'. Regular traffic began on 1 July and was surprisingly heavy. In 1877, going ahead of the rest of the story, the LSWR applied to the Railway Commissioners for a third rail to be laid between Lydford, reached from Okehampton in 1874, and Launceston. The South Devon claimed that 65 per cent of Launceston's trade was with Plymouth, and defeated the application. But soon, of course, the LSWR built a direct line between Okehampton and Launceston.

ENTER THE LSWR

The 1850s and 1860s saw both expansion and consolidation in the Three Towns, Plymouth gaining most. Yet the railway lacked enterprise: until 1871 there were but two stations to serve nearly 100,000 people. By the early 1870s the paucity and slowness of the trains was retarding the area's development; immediately before the LSWR entered the scene, only 20 week-day trains left in winter, nine for the South Devon line, five for the Tavistock branch, five to Truro and Falmouth, and one to Liskeard. Many passengers as well as goods were still going by sea.

When the LSWR's narrow trains arrived, on 18 May 1876, they were greeted more enthusiastically than had been Plymouth's first on the South Devon many years ago, though the mere sight of an engine was then a novelty to many. Actually 18 May was more Devonport's day than Plymouth's. By means of a third rail the LSWR shared the single broad-gauge line of the Tavistock & Launceston route from Lydford to Marsh Mills, and still kept to the GWR through Plymouth – a new joint station at North Road was not ready for the opening – only branching on to its own line again near Devonport, where the handsome terminus was a fillip to the borough's pride. Millbay was by-passed by the opening of the Cornwall loop. Plans for an independent line beyond Marsh Mills had been defeated in both 1872 and 1873.

As both broad and narrow traffic grew, the working of the

mixed gauge single line between Lydford and Marsh Mills became complex. For a short time there was a mineral branch from Shaugh Bridge to a quarry at the Dewerstone Rock, and on 11 August 1883 part of the line was further enlivened by the opening of the narrow gauge Princetown Railway, a GWR subsidiary, whose trains at first ran through from the junction at Yelverton to Horrabridge. The new branch was 10½ miles long and much of its route followed the old Plymouth & Dartmoor, of which this section had been bought for £22,000. In the early years the branch was chiefly concerned with taking traffic (convicts, officers and goods) to Dartmoor Prison, but King Tor quarry still produced some granite. Later, excursionists were most numerous among the passengers, though not numerous enough to prevent closure. On the last day, 3 March 1956, two engines and six coaches were needed to accommodate the sightseers. Not that they saw much: Dartmoor shrouded itself in fog. Those of us who loved this lonely railway by-way were naturally sad; but the earthworks and granite cuttings are relics not easily eradicated.

Throughout the 1880s the LSWR was fretting for an independent route to the Three Towns, the GWR to rid itself of its peculiar gauge. Meanwhile neither Company heeded lesser things, the train service not improving as much as the extra traffic of the period warranted. But after the opening of the new Lydford–Devonport line on 2 June 1890 and the gauge change in May 1892, at last the scene was set for unparalleled competition and initiative.

The new Lydford–Devonport line belonged to the Plymouth, Devonport & South Western Junction Railway Company, formed in 1883 with extravagant ambitions which seem to have worried Waterloo. However, though holding a very small proportion of the capital, which eventually totalled over £1,000,000, the LSWR worked the line from the start. The cost was high because the 22½ miles of double track included numerous engineering works, such as the Tavy and other long bridges and three tunnels, two of them under Devonport to enable the LSWR terminus to be reached and turned into a through station. The small Company remained reasonably prosperous till its absorption by the LSWR in 1922; as well as the main line it owned the Callington branch, which it worked itself.

The Plymouth, Devonport & South Western Junction is often quoted as an example of needless duplication. Had the GWR been improved between Lydford and Plymouth, it could easily have handled all the traffic, though the new line did give the Tamar Valley its first main railway. Bere Alston and Bere Ferrers, then isolated by their peninsular position, soon developed, and market gardening was much helped. The coming of trains spelt doom for boatmen, the regular services up and down the river and several ferries across it quickly succumbing.

At first the trains over the new line terminated at North Road, but in 1891 they were extended to Friary, a new four-platform terminus in eastern Plymouth. This district, incidentally, had suddenly become the centre of operations for the Plymouth & Dartmoor Railway Company, reawakened to act as an LSWR agent. The Turnchapel branch was built under its auspices, and on 22 June 1888 an Act was obtained for the South Hams Railway, to leave the Turnchapel branch at Plymstock and go to Modbury. The GWR feared its extension to Torbay and Exeter. By an 1894 Act, however, the rights for the Plymstock–Yealmpton section were transferred to the GWR, and for the Yealmpton–Modbury part to the LSWR, each to have running powers over the other and act as mutual checks. Soon the LSWR dropped interest, neither building the extension to Modbury nor exercising running powers to Yealmpton. The GWR gradually increased the Yealmpton service (which ran from Millbay) and the LSWR that of the Turnchapel branch, until Plymstock had over 60 daily trains. Bus competition made early inroads on the traffic: dates are given in Appendix II.

1900 ONWARDS

It was the LSWR on the Plymouth, Devonport & South Western Junction line which first made special provision for main-line suburban traffic – in the 1890s. Several locals ran from Devonport to St Budeaux, one of them timed progressively earlier as the autumn evenings darkened. The Dockyard could then work only in daylight. At the turn of the century all the GWR offered exclusively for local travellers were two trains from Millbay to Plympton and another two to Saltash. Most Saltash people crossed the river by steam ferry to St Budeaux or Devonport.

In 1903 about 100 week-day trains passed through North Road; four years later there were 250. Such was the GWR's sudden awakening. The suburban services were run by the new steam rail cars, the engine built into the coach; the first went into service in Plymouth in 1904, when several halts were opened and substantial engineering and signalling works undertaken. Fares were drastically reduced: it now cost 6d return from Saltash to Millbay. Hundreds of workers saved an hour or more a day going to and from the Dockyard or the docks, and greater leisure coupled with better evening travel facilities resulted in more social life. Many people moved to the suburbs: St Budeaux, included in Plymouth's boundary in 1896, grew most rapidly, but all the halts had houses built within walking distance. In 1906 the Plymouth, Devonport & South Western also opened halts, increased the St Budeaux–Devonport local service and extended it to Friary.

Between 1881 (when the GWR still had no station between Devonport and Saltash) and 1914 (when during the summer over 40 down Monday-to-Friday trains, mainly rail motors, between them made a total of about 120 calls at the four new stopping places) the population of the Three Towns rose from 138,000 to 214,000. Outside the boundary of the now united Plymouth, Saltash (served by up to 50 trains a day from Plymouth) and Plympton (with up to 25) grew even more rapidly – and so did the villages served by the Yealmpton and Turnchapel lines. In the central areas population density thinned, so that the total acreage devoted to housing must have at least doubled between 1881 and 1914. Although slow on the uptake, the railways paved the way to better living. They were not to reap the benefit for long, however. Once the tide of expansion into the surrounding country had begun, it rushed on: soon residential areas were being developed miles from a station. Moreover, when the Three Towns had amalgamated in 1914 their hitherto unconnected tramway systems had been linked. Several halts soon closed, but it was not until buses were well established in the 1920s that loss of traffic became serious. Rail motors were withdrawn from the Plympton run in 1930. The Saltash service survived till 1970 though substantially reduced following the opening in 1961 of the one-span road bridge slightly to the north of Brunel's masterpiece. Before the war a difficulty was that while some trains went to Millbay, others had to run to North Road (which replaced Millbay and Mutley in importance in about 1920) for main-line connections. This division did not suit Saltash people. Now North Road is the only station in Plymouth proper.

By improving the service and offering cheap fares the GWR kept the Plymouth–Tavistock section of its Launceston branch busy until 1939, carrying up to 20,000 excursionists on bank holidays. After the war the figure seldom reached 2,000 and in the later 1950s many winter trains carried fewer than ten passengers. The entire 32 miles from Tavistock Junction to Launceston were closed to passengers (for goods details see Appendix II) from 31 December 1962. Closing night, 29 December, coincided with West Devon's worst blizzard of the century. The last trains in each direction were unable to run and the previous two had to be abandoned *en route*. Before this final decline of traffic the Tavistock branch, like all other lines leading into Plymouth, had an extremely busy wartime spell. Buses were infrequent, and thousands of Plymothians temporarily moved to neighbouring towns and villages. Goods traffic, too, was enormously increased. Much railway property was damaged in air raids and frequently through traffic had to be diverted, though no single incident stands out. By contrast, the 1914–18 war made surprisingly little difference to Plymouth, apart from the embarkation and disembarkation of troops, usually carried to or from the city by the LSWR.

Nearly all Plymouth's harbours were directly or indirectly developed to their present form by the railways. Business at the quays flourishes, at the expense of smaller harbours elsewhere in the West. Like other places, too, Plymouth receives an increasing proportion of its coal by rail from the collieries. Other coal, however, is still imported at Sutton or Cattewater harbours and distributed by rail. The passenger side of the port has special railway interest: it was an ocean special which first reached 100 m.p.h., while another was involved in one of the worst accidents in southern England. Much of the exciting story of the ocean traffic belongs to Chapter XI, and for the sake of convenience is all included there.

Today the emphasis in Plymouth as elsewhere in the West lies on long-distance traffic. Nostalgia is easy, of course: before dawn the halts now in the last stages of decay were once thronged with

Dockyard workers; between eight and nine o'clock came the school-children; noon was the shopping rush-hour, with perambulators parked down the centre of the auto-trailers; and best remembered of all, perhaps, are the 'fish-and-chip specials' out of Millbay late at night, especially on Saturdays. The journey formed part of the evening's enjoyment far more than it does today. But to look too long at the halts and other ruins of a bygone era is wrong: times inevitably change, and it is difficult to see how a useful suburban service could be made to pay on most lines today. The only regret about suburban services must be that they were not started earlier, giving them a longer heyday before the bus arrived. What is sad about the whole Plymouth railway story, indeed, is that opportunities were taken slowly – at least until the turn of the century. However, in the new North Road, with its skyscraper block, its Brunel Bar, and colour-light signalling, the city has a station infinitely superior to anything it has previously possessed. The service to London is the quickest and most frequent ever. There also remains an excellent service into Cornwall though alas many fewer people make 'local' journeys by train even if they are journeys of 50 or 80 miles along the main line whose expresses thus tend to become increasingly empty the further west they are from London.

In September 1964 the war-time connection between the two lines at St Budeaux was brought into permanent use and the former Southern line through Devonport closed. Fuller details of Plymouth's changing railway system are given in Appendix II.

TAMAR VALLEY

Mines in the Gunnislake district and quarries near Kit Hill had long needed better transport when the East Cornwall Mineral Railway, opened on 7 May 1872, came to the rescue. Seven miles long and of 3 ft 6 in. gauge, it began on Calstock Quay and rose by a cable-operated incline to the cliffs above, beyond which locomotives were used to the terminus at Kelly Bray, a mile from Callington and 640 ft above sea-level. During its early years it prospered, the first half year of 1876, for example, producing a profit of £1,140 out of £2,547 receipts. Soon the railway proposed to link itself with a tramway from the Devon Great Consols Mine to Morwellham Quay on the Tamar. But bad times came and in 1891 the existing line had to be taken over by the Plymouth, Devonport & South Western Junction. Under Light Railway Orders of 1900 and 1905, its gauge was converted to standard and it was extended across the slender 129-ft-high stone viaduct to join the main line at Bere Alston. From the start traffic was heavy, but so were expenses. An original 1 per cent dividend increased to 3½ per cent in 1913, when 112,639 passengers were

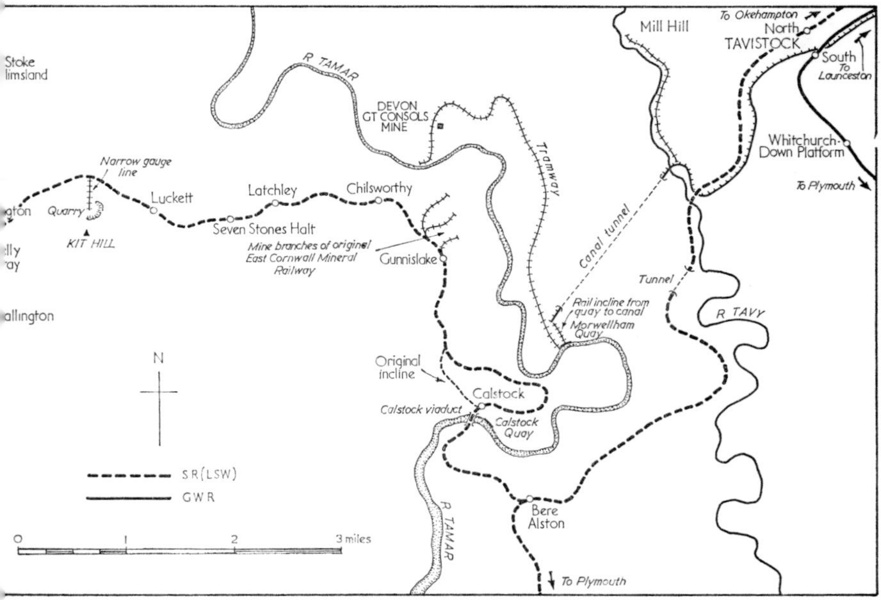

Tamar Valley

carried. Between the wars mineral traffic almost disappeared, but market-garden produce increased. In a typical year in the 1950s 400 vans and 50,000 packages of flowers and 300 vans and 66,000 packages of fruit were dispatched from Calstock and Gunnislake. Though among the least-known by-ways, the branch was full of character, giving fine views of the Tamar and its forgotten quays and thickly wooded banks, its crumbling mine stacks and its patchwork of small and keenly cultivated market gardens.

Closure beyond Gunnislake took place from 7 November 1966, since when Plymouth–Gunnislake with reversal at Bere Alston has been retained as a self-contained branch whose closure even today would cause hardship since road routes are much longer than the railways with the bridge over the mouth of the Tavy and the viaduct at Calstock.

The Devon Great Consols tramway, five miles long, with 3 locomotives and 60 waggons, connected the Tamar Valley's richest mines with Morwellham Quay. The quay had an extensive siding system, one of the lines running up an inclined plane to the terminus of the Tavistock Canal (whose Mill Hill branch, serving a short-lived quarry, was in 1844 replaced by a tramway). A narrow-gauge tramway not reaching Morwellham Quay ran when part of Devon Great Consols was temporarily reopened in 1919. Morwellham Quay is now an industrial archaeology centre.

CHAPTER VII

Cornwall

THE CORNWALL RAILWAY

> Hereafter, when trains moving in an opposite direction are approaching each other on separate lines, guards and drivers will be required to bring their separate trains to a dead halt before the point of meeting and be very careful not to proceed until each train has passed the other.

The quotation from the official rule book cannot be resisted. It is one of many details which make the Cornwall Railway one of the most interesting in the West, perhaps second only to the Somerset & Dorset. But too much must not be read into the rule, for the Cornwall sank into chaos less frequently than many other lines of the date – and certainly the general standard of efficiency was well above that of the Somerset & Dorset. The Cornwall's interest lies chiefly in its struggle against economics and geography. Its shortcomings, which admittedy included rather too many mishaps, must inevitably occupy much of this account; but it would be wrong to minimize its part in developing Cornwall, an aspect considered at the end of the section.

As early as 1835 a Bill for a Central route from Exeter to Falmouth passed the Commons but was rejected by the Lords. In 1839 both this (there were to be 10½ miles of 1 in 38 incline worked by water power and ropes) and a Coastal scheme were discussed. In 1843 a delegation from a county investigating committee was told by the GWR that the South Devon Railway could not be extended beyond Plymouth, but that the broad-gauge Companies might help a separate Cornish concern to develop a Coastal route.

Tories backed the Coastal, now generally referred to as the Cornwall Railway, and Liberals the Central. Argument increased when the LSWR openly supported the latter. Throughout 1844 the

West Briton flourished on the battling. One week it carried an announcement boasting that the LSWR's subsidiary Devon & Cornwall Central, as it had become with a proposed capital of £1,500,000, was 'not liable to interruption by any arm of the sea', while a Padstow branch would open up a shorter route to Ireland. The next issue brought a front-page 'Exposure of the Fallacies contained in the statement on the superiority of the Central Line'.

Both Cornwall and Central were ready for Parliament in 1845 – but the Cornwall only just so: see Appendix III. The Cornwall alone passed the Board of Trade scrutiny. The Board agreed that the 26 miles to be graded at 1 in 60 might not be objectionable as the atmospheric system was to be used, and a proposed ferry instead of a bridge over the Tamar was 'no insurmountable impediment'.

J. T. Treffry, chairman of the Cornwall, wrote (in a letter now in the Treffry Estate papers): 'Our railway prospects are much mended by the decision of the Board of Trade. We have, however, *between ourselves* many difficulties to get over from the absence and inattention of our engineer. I have heavy complaints against him.' As well he might. Captain W. S. Moorsom coolly charged £27 for a single journey from London, and in his supreme self-confidence took no pains to placate landowners. Anyway, the Lords did not view his plans as favourably as had the Board of Trade, and a new survey for another Bill was the work of Brunel. This Bill became an Act on 3 August 1846, allowing £1,600,000 capital for a double Plymouth–Falmouth line (including a bridge over the Tamar) and several branches. Between them the GWR, Bristol & Exeter and South Devon – the Associated Companies – were to contribute £337,000. Among existing lines the Company was empowered to buy was the Bodmin & Wadebridge; but before the Act was passed the LSWR had already acquired it on behalf of the Devon & Cornwall Central, which made another unsuccessful bid for life in 1847.

The Cornwall's first sod was cut 'in a spiritless manner' near Truro in August 1847. Riots followed, it being feared that the influx of navvies would mean insufficient food for everyone. But the navvies were quickly dismissed, the Company falling asleep during the reaction to the Railway Mania. An attempt by Brunel

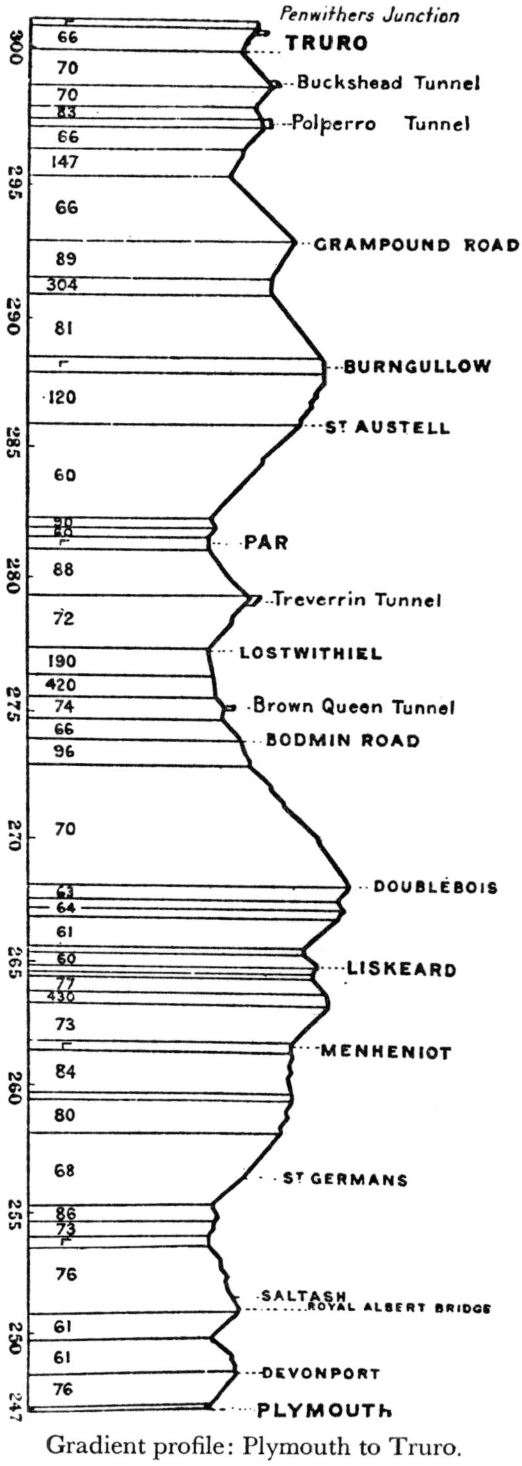

Gradient profile: Plymouth to Truro.

to awaken it in 1851 was fruitless; but in April 1852 shareholders were called to Truro to hear that the Board of Trade had granted a warrant for the abandonment of all branches (except for Bodmin Road–Bodmin; even this was discreetly forgotten) and for a reduction of capital to £1,125,060, the main line's track to be single only. Shares, hitherto of £50 each, were converted to £20.

At last work restarted; but during 1852 9,583 shares were forfeited and could not be disposed of to new holders. Then the contractor on the Truro–Falmouth section failed and it was decided to leave the works half-finished and concentrate resources on Plymouth–Truro. These resources were limited in the extreme, for by 1854 no fewer than 29,229 of the total of 56,253 shares had been forfeited. (34,502 were forfeited by 1859.)

After a futile attempt to rearrange the capital, an urgent appeal was sent to the Associated Companies. Leisurely negotiations resulted in their agreeing, in June 1855, to lease the line jointly for seven years from its opening between Plymouth and Truro, or until all expenses incurred by them had been recouped. Meanwhile they guaranteed interest on a £375,000 debenture issue: this enabled the Cornwall Company to let the contracts for the one section between Plymouth and Truro not yet started. As time went on, this figure of £375,000 guaranteed fixed interest stock was vastly increased, the first addition being £40,000 of preference shares mainly for the purchase of rolling stock issued in 1857.

As stated in the last chapter, the Cornwall Railway, including the *pièce de résistance*, the Royal Albert Bridge, was opened to public traffic from Plymouth to Truro on 4 May 1859. Unofficially, passengers seem to have been carried on the western end of the line during the preceding months. Locomotives for this part had been brought by sea to Par. From the start of regular traffic the line was worked by a Joint Committee, with representatives of the Associated Companies and of the Cornwall. Goods began on 3 October, but few facilities were ready.

A journey on the Cornwall line is a geography-history lesson in itself, the scenery ranging from sea and estuaries to luxuriant fresh-water river valleys and bleak hills. Some of the least spoilt villages in the county can be seen, and the quaint almshouses at St Germans, historic Restormel Castle near Lostwithiel, countless minestacks and china-clay dries and tips—not to forget the magnificent bird's-eye view of Truro and its cathedral just before the station. That the trip is so rewarding is largely because there are relatively few deep cuttings but numerous embankments and viaducts – originally 34 of the latter, 31 of them between Saltash and Truro, with a combined length of four miles. Except for those across tidal creeks east of St Germans, they had masonry piers up to 35 feet below rail level. The longitudinal beams were supported by a fan of four timbers radiating from the top of each of the piers, spaced 66 ft apart. St Pinnock (in the Glyn Valley) and Liskeard Viaducts were 151 and 150 ft high respectively; Truro was the longest, 443 yds. Though the best kyanised yellow pine was used, the viaducts were cheap to build; but they were hopelessly expensive to maintain, particularly those on a curve or an S-curve. Replacement by masonry or iron structures, or in some cases by embankments, began in the 1870s. Other engineering works included several tunnels, but none longer than the 581-yd Polperro. The diagram on page 131 shows the line's switchback character, though only chief gradient changes are indicated.

An Act of 1861 renewed powers for the Falmouth extension (see branch lines) and authorized a 1,000 year lease to the Associated Companies, the new Joint Committee consisting of four Cornwall, three South Devon, three Bristol & Exeter and two GWR members. The arrangement was that surplus revenue, after working and all fixed charges had been met, should be used firstly for payment of guaranteed dividends and secondly to recoup the Associated Companies for past losses. Had there ever been a surplus after that, it would have gone to the Cornwall as rental; but by 1875 the Associated Companies had been forced to pay £137,325 of the guaranteed dividends out of their own funds.

Reasons for the paucity of traffic in the early years may be left till later; traffic figures for 1861 will be found in Appendix I, which also lists the limited rolling stock. For convenience between the lessees, locomotives were provided by the South Devon, or at first by that line's locomotive contractor. Track maintenance was appalling, trains rolling violently even at moderate speeds: higher speeds were anyway seldom obtained. This, the tenuous appearance of the viaducts, and the number of mishaps discouraged much local patronage.

The first accident occurred only two days after the opening;

just when the public's confidence needed winning, the engine of a passenger train plunged off the rails at the approach to a viaduct near St Germans and embedded itself funnel downwards in the mud. Three railwaymen were killed. For many years thereafter the derailment of odd coaches owing to the poor track could legitimately be described as commonplace. Crossing movements were controlled by telegraph from the beginning, but distants were the only signals. Trains were started from stations by word of mouth. The worst accident, a head-on crash in a cutting between a goods and the down mail, in 1873, resulted from the guards of both goods trains at Menheniot station being called Richard. The wrong one acted on the signalman's 'Right away, Dick'. The previous year a head-on crash had been narrowly avoided by the driver of a passenger train between Par and St Austell spotting in good time a china-clay train coming down the hill in the opposite direction out of control. With the greatest alacrity he flung his locomotive into reverse, and to the consternation of passengers retreated the way he had come. Following an accident at Truro in 1861, the Board of Trade instructed that interlocking should be introduced forthwith at all stations west of Exeter, but little was done (either on the Cornwall or the South Devon) until the 1870s and 1880s.

The timetable (see page 151) was little altered for many years. Until 1876 third-class passengers were carried by all services at 1d a mile, but thereafter they were treated as brusquely as on the rest of the GWR. For from 1876, owing to the amalgamation of the Bristol & Exeter and the South Devon, the GWR became sole lessee of the Cornwall, which was virtually included in its own system: occupying eight of the twelve seats (including those previously held by the Bristol & Exeter and South Devon), it naturally dictated the policy of the Joint Committee. But the Cornwall Company, still holding four seats, not only continued in existence for another thirteen years but succeeded in being extremely troublesome.

In fact the Company now represented the interests of the holders of only a fraction of the capital. Out of the total of £1,833,000, no less than £1,258,000 was in debenture and preference shares on which the GWR guaranteed the interest. Nearly a third of the ordinary capital was also held by the GWR itself; but some of the remaining £373,000 belonged to speculators, who had bought shares cheaply and were intent on raising their value. They forced the Cornwall Board to reject cash offers made by the GWR – £3 10s per £20 share, increased to £4 10s. In 1882 the railway at last made profit more than enough to meet the guaranteed dividends, but before any dividend could be paid on the ordinary shares, the arrears owing to the GWR – now totalling some £230,000 – had to be cleared. What the speculators now did was to make the Board peremptorily announce that in future it, and not the Joint Committee, would decide when to replace viaducts: the less the expenditure on such works, the sooner would there be a dividend. Two viaducts which the Joint Committee intended rebuilding wide enough eventually to take two narrow-gauge lines, were made the particular subject of dispute which went to arbitration.

The arbitrator, Sir Arthur Hobhouse, decided that the Joint Committee might rebuild viaducts where prudent for the working of the line. But as the lease stipulated that the gauge should be broad, if the disputed viaducts were rebuilt they would have to be of broad-gauge width, whether for single or double track. As nobody in his right mind would have built new works for the long-doomed broad gauge in 1884, the speculators won the day, at least in practice. After all, however, the Company gave its consent for the two viaducts to be made of double narrow-track width, though it was agreed to rebuild no more for five years. Before the period expired, the Company had gone. After a cash offer of £8 per £20 share had been accepted, it was dissolved on 1 July 1889. Some of the speculators must have made a profit of 250 per cent; the local people who had bought shares at the beginning recovered less than half their capital after more than a generation without a dividend.

Even today Cornwall is Britain's most individual county, but differences from the 'mainland' are nothing to what they were before the opening of the Royal Albert Bridge in 1859. The bridge was supposed to let in the Devil. What in fact trickled in were all kinds of national unifying influences – to the true Cornishman, of course, these were the very Devil. But the old order disintegrated only slowly, and it was largely for this reason that the Cornwall Railway was so poor a proposition financially. For instance, Cornishmen were reluctant to accept mass-produced Midland goods, while only a few were persuaded to venture as far afield in their own country even as Plymouth – though many had gone straight from Cornwall to seek employment in mines all over the world.

It was, indeed, only after thousands of miners had been forced willy-nilly to emigrate in the depression of the 1870s that tourism and agriculture increased: both helped the Cornwall Railway far more than the two traditional sources of livelihood, mining and fishing. Nearly all minerals continued to leave the county by sea, being taken to the ports by a series of independent railways established well before 1859. With the exception of the West Cornwall, none of these early lines immediately obtained connections with the Cornwall Railway, remaining isolated concerns – in some cases handling greater tonnages than did the east–west trunk route itself. The richest mining area was in any case in the far West, beyond the Cornwall, which for much of its way crossed a thinly populated no-man's-land. Originally only the prospect of increasing Falmouth's importance as a port had persuaded some of the promoters to risk their money. Had it been realized that the departure of the lucrative packet business in 1844 was final, the delay in linking Cornwall with the rest of the kingdom would almost certainly have been yet greater.

THE WEST CORNWALL RAILWAY

In 1801 there were 75 mines in Cornwall, employing about 16,000 people. In 1862 there were 340 mines, employing 50,000 people. The expansion was due to three chief causes: higher copper prices on the world market, the adoption of steam pumping machinery to enable the working of deep lodes, and better transport facilities. The latter included extensive underground tramway systems, from about 1824, as well as the mineral lines to the ports. The West Cornwall was perhaps the most important of the pioneers – and had the distinction of being the first part of the GWR Paddington–Penzance trunk route to be constructed. Of still earlier origin, however, were the Poldice Tramway (1809 or 1810), Redruth and Chasewater (1826), Pentewan (1830) and Bodmin & Wadebridge (1834), which we shall deal with in due course.

Nearly all the lines built before 1859 were partially, if not chiefly, for copper, Cornwall's undisputed king throughout the first half of the century. It made the county wealthy, ill-distributed though the gains might be. The rich led a life as gay as anywhere in the provinces; the other side of the picture may be gauged from the fact that out of every 100 deaths in mining parishes at one date, 55 were of children under the age of five. The lives of miners were proverbially short, their ends often violent. But mining was not merely a job: it was a way of life. Its attraction was irresistible, largely because even the humblest miner had the remote chance of a lucky find. Those who did not mine were mainly fishermen, of course – gamblers in another element. Agriculture was too humdrum and often neglected, with the result that fear of food shortage was sometimes acute, as demonstrated by the rioting after the Cornwall Railway's first sod had been cut.

But local prices were low, and after 1859 butchers, for instance, soon found it worthwhile to dispatch meat to London. In 1864 the *West Briton* complained that the higher prices, which Cornishmen could ill afford, were 'not to be wondered at when we are informed that not less than 600 head of sheep are sent away from the county every week'. Nor was it only meat that disappeared. As early as 1861 the Cornwall Railway's directors had 'observed with pleasure' that 1,063 tons of fish, 1,787 of potatoes, and 867 of broccoli had been dispatched in the season; in 1862, 466 tons of new potatoes and 212 of mackerel left from West Cornwall in a season. The potatoes and broccoli were grown chiefly for 'export' (a small quantity had travelled by the Hayle–Bristol boat service in pre-railway days); but fish was almost the Cornish staple diet, and great hardship followed when the old-established pattern of trade was interrupted. Fish landed one day could be on sale in London the next, and this encouraged a fleet of drifters to local waters, to the detriment of the traditional seining. At Newlyn, where trade developed dramatically after 1859 owing to the proximity of Penzance station (lesser ports immediately declined), some of the once-prosperous self-employed fishermen ended as porters employed by the drifter-owners.

Just as the railway's unfortunate influence on general fishing was being felt, the pilchards which had played a vital part in the

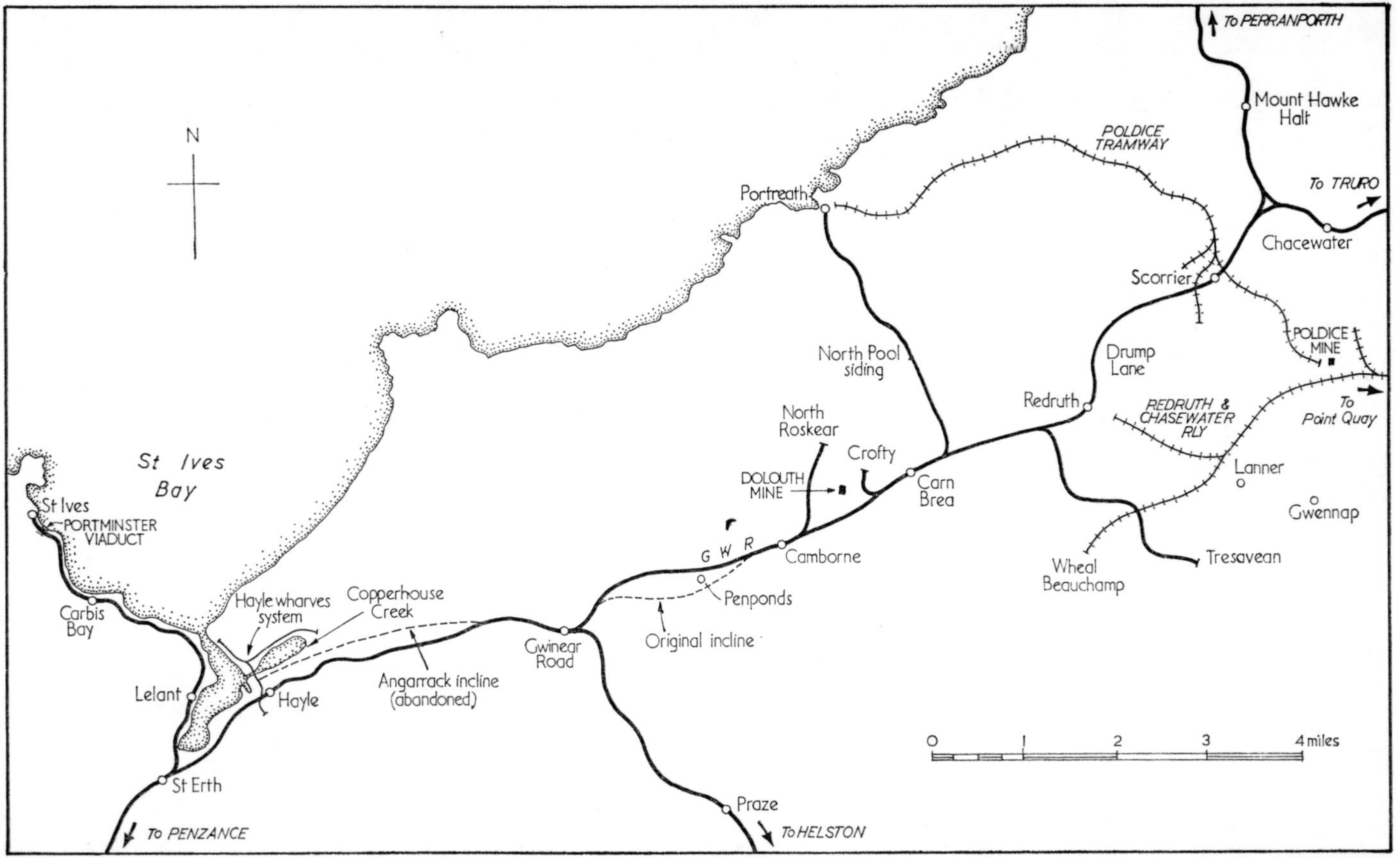

West Cornwall.

economy disappeared from the Cornish coasts – and copper production began declining. The 1870s were even worse. The bottom fell out of the copper market. The population, which had risen steadily since 1800, dropped precipitously, a quarter of the miners leaving in the decade. New sources of livelihood, made possible by the railway, provided little help before about 1890, and first we must proceed with an account of the West Cornwall and its forerunner, the Hayle Railway.

Hayle was the chief mine port, seat of a growing foundry and base for hundreds of packhorses and mules when in 1834 the narrow line named after it was authorized, with a capital of £64,000. Opened in stages beginning on 23 December 1837, its total mileage was over 17, the main line, from Hayle to Tresavean (Gwennap), being 12 miles. It left Hayle along the northern shore of Copperhouse Creek and rose by 1 in 10 Angarrack Incline to join the course of the present main line west of Gwinear Road. There was a second incline at Penponds, a third near Redruth, and a fourth down to the harbour on the Portreath branch. Other branches ran to Hayle wharves, Crofty and Roskear mines, and Redruth, just by-passed by the main line.

Locomotives were used except at Hayle and Portreath and on the inclines leading from them. Eventually they also ran to Hayle, attached to the ropes on Angarrack Incline. A passenger service was started on 23 May 1843 at the instigation of William Crotch, proprietor of a Hayle hotel. His 'neat and commodious omnibuses' made two daily journeys between Hayle and Redruth, also conveying merchandise: 'Every care will be taken to secure speedy and safe delivery at the lowest rate of charge ever known to the public.' Apparently Crotch soon relinquished his interest, the income table for six months of 1849 suggesting the railway was working all traffic itself:

Copper ore	£7,652	11s	4d
Coal	£3,513	1s	10d
Sand	£ 269	9s	6d
General Merchandise	£ 546	11s	10d
57,018 Passengers	£1,486	0s	4d

The West Cornwall Railway was authorized on 3 August 1846, with £500,000 capital in £20 shares, in order to rebuild the

Hayle Railway and extend it to Truro and Penzance. (In 1845 a first Bill had been rejected on the grounds that the district merited a better scheme than that proposed. As on the Cornwall Railway, Brunel replaced Captain Moorsom as engineer.) The Act compelled an application in the next session for power to provide a public wharf at Hayle and to construct a branch on the west side of the estuary to Norwayman's Wharf. The application was duly made, but the wharf was rejected on the grounds that a railway should not be a trading concern, and the branch because the Admiralty objected to the crossing of the river!

Reaction to the Railway Mania was felt here as elsewhere, 7,635 out of 21,000 shares being forfeited, and some resold at discounts of up to £6. The gauge was to be broad, but as the Cornwall Railway was making no progress the Company asked to retain the existing narrow gauge. This was allowed by an Act of 1850 which stipulated, however, that works should be made wide enough for the broad gauge and that this should be laid within six months of notice from a connecting broad-gauge company.

There was further delay because Brunel had decided to use Barlow rails, shaped like an upturned flat-bottomed V, their two ends embedded deep in ballast. Only recently patented, their manufacture gave considerable trouble, and then one of the first cargoes was sunk on the crossing from South Wales. Public traffic (three daily passenger trains) began from Penzance to Redruth on 11 March 1852. An extension to Truro Road was opened on 25 August. From 16 April 1855 the Truro terminus was beside the river at Newham, where a wharf was provided.

The new line by-passed the Hayle Railway's Penponds and Angarrack Inclines by viaducts and crossed Hayle by the same means. Altogether there were nine viaducts, six of a similar design to those on the Cornwall Railway but of lighter construction, with three instead of four supports between the tops of the piers and the longitudinal beams. Penzance Viaduct, one of three with straightforward vertical timber supports, was partially demolished by the sea soon after the opening and was practically destroyed in 1868. A new viaduct was opened in 1871 but was replaced by an embankment in 1921. The West Cornwall's chief gradients lead up from Hayle and Truro respectively to the plateau section, roughly between Camborne and Scorrier.

The railway was not prosperous, mainly because the principal mines served were among the first to show declining copper output. Because of the line's isolated nature, general traffic was inevitably light, though occasional excursions must have been profitable enough. The first of these in 1852, took the Redruth and Camborne Temperance Societies to Hayle. Three squat locomotives struggled valiantly with 76 mineral trucks filled with teetotallers singing:

Happy Camborne, happy Camborne,
Where the railway is so near;
And the engine shows how water
Can accomplish more than beer.

On the return, the locomotives ran out of steam and the procession halted opposite an orchard. In the words of one chronicler:

> It may have been their extreme anxiety to take measures against such an intoxicating beverage as cider, but at all events that army of teetotallers swarmed down from the trucks and up the apple trees until the orchard resembled the famous cupboard of Mother Hubbard.

Though the Company was willing enough to take advantage of the Cornishman's temperance enthusiasm, his love of bright clothing was quashed by a rule that no employee should wear 'apparel of a Red or Pink color'. Another rule was: 'The Guard is to signal the Engine Driver if he is going Too Fast or Too Slow.' The Company was handsomely treated by the Press as free passes were distributed liberally.

At the opening the chairman had commented on the 'evenness, firmness and ease with which the carriages passed over the line', but the Barlow rails caused numerous derailments. The Board of Trade made various suggestions, and in August 1857 the directors reported: 'The improvement of the Line is not only evident to those travelling over it, but the damage to Rolling Stock is much diminished.' But Barlow rails had gradually to be replaced.

From 11 May 1859 all but one daily train each way used the Cornwall Railway's Truro station (which became joint property), reached by a new line from Penwithers Junction. The Newham line became a branch. It was not until 1864 that the Cornwall gave notice for broad-gauge rails to be laid to Penzance.

The Company was in no fit state to meet the demand. It had paid its first dividend in 1852 – 3s 3d per £20 share, for the second half of the year, but this was reduced to 2s for the next three and a half years, and then nothing. An Act of 1865 gave the Company power to sell or lease itself to one or more companies, broad or narrow. (The latter would have been the Cornwall Central: see Chapter V.) The broad-gauge Associated Companies took over maintenance and working from 1 July 1865, and from 1 January next the undertaking was transferred to them absolutely in return for an annual rent charge beginning at £21,000 and rising to £25,000 in 1873. This was enough to pay the 5 per cent interest on preference shares and 2¼ on the ordinary shares. The West Cornwall Company was dissolved in 1878.

Unlike the Cornwall Joint Committee, of course, the West Cornwall Railway Committee consisted entirely of the representatives of the GWR, Bristol & Exeter and South Devon, whose first task was to lay the broad-gauge rail. After some viaduct reconstruction (though the West Cornwall's new works were built wide enough for the broad gauge, several of the old Hayle Railway's viaducts had been retained unaltered), this was ready for the first broad-gauge goods to Penzance on 6 November 1866. Broad passenger trains did not start until 1 March 1867, the Board of Trade refusing to sanction the works until the Associated Companies guaranteed that all stations would be served by narrow trains, brought to the platform face from one side of the broad gauge to the other by means of fixed facing points. Thereafter four of the five daily passenger trains (two with through coaches from Paddington) were broad, but most goods traffic continued narrow. From 1871 there was a daily goods of the waggons of both gauges. The South Devon provided broad-gauge locomotives (as in the case of the Cornwall Joint Committee, a transfer of some machines to the West Cornwall Committee was merely a paper transaction) and converted several of the narrow-gauge engines for the broad. The remaining narrow engines were based on Carn Brea, headquarters of the mineral traffic. In 1862 the West Cornwall directors had announced a decrease in the total number of passengers of no less than 39,105, coupled with an increase in passenger receipts of £514, 'which arises from the discontinuance of the issue of Third Class tickets by all trains. These tickets are now issued by two trains only' – an arrangement which the Committee maintained until 1876, when the railway became exclusively the GWR's. From then till 1899, when the Cornwall Railway followed suit, there were few changes.

SALTASH–PENZANCE AFTER 1899

If the main line had caused more embarrassment than it was worth in its first years, it was vital to Cornwall's economic survival in the decades which now followed. But before it could play its part effectively there had to be vast improvements. A glance from the trains shows deviations from the old route at dozens of points, most of them adopted when the track was doubled. Work started immediately after the abolition of the broad gauge in 1892 and continued on one section or another for ten years. The opening of the new Truro viaduct on 14 February 1903 finished the doubling between St Germans and Penwithers Junction. Between St Germans and Saltash the original route was abandoned for a better one inland, involving three instead of five viaducts, but also the 450-yd Shillingham Tunnel. It was used by passengers from 1 June 1908. Progress on the West Cornwall was slower, and only on 13 April 1930 was the last section (Scorrier-Drump Lane) ready, completing the double line from Paddington to Penzance save for the Royal Albert Bridge. (To save the cost of rebuilding, the track is now singled over two adjacent viaducts between Liskeard and Bodmin Road.)

Passenger traffic must almost have trebled between 1890 and 1910. Cornwall had no old-established resorts and before 1890 visitors were remarkably few; but as in South Devon the middle- and lower-class holiday business now grew rapidly. Local patronage increased, particularly in West Cornwall where rail motors began augmenting trains in 1905 when the *Cornish Riviera Express* was introduced. Its speed through Cornwall was little increased until the 1970s, the introduction of 4-6-0s some years later merely allowing greater weights: platforms at nearly all intermediate stations had to be lengthened. Trains steadily became more frequent too, with greater variations between the winter and summer services. Many minor improvements were made to stations between the wars; the chief scheme was that at Penzance, where in 1937 the old unsatisfactory terminus was replaced by the present workable four-platform arrangement, though the ugly overall wooden roof was retained. Siding and locomotive accommodation between Penzance and Marazion was augmented on several occasions.

Today tourism, as mining in the past, heavily overweighs Cornwall's economy. The lack of balance is undesirable but inevitable (isolation puts industry at a disadvantage), and the value of the GWR's energetic campaign in placing the county on the holiday map cannot be overestimated. However, the increase in visitors since the war has far exceeded any rise in GWR days, and though a large proportion of tourists now go by road, the railway's business continued to expand until 1958. Altogether something like 200,000 long-distance visitors then crossed by the Royal Albert Bridge in each direction during the season; on the busiest day of all the total, including local passengers, was probably over 25,000 each way. Although Par has become a camping centre (the station serves a wide area as well as being junction for Newquay), Penzance is the only full-scale resort on the main line, receiving up to 2,500 long-distance passengers on peak Saturdays in the 1950s.

Cornwall's season is shorter and sharper than Devon's. This, coupled with the heavy gradients, made working highly expensive in steam days. Trains of more than ten coaches had to be double-headed. Moreover, the majority of passengers were for branch-line destinations, and many expresses were – and still are today – almost empty before they reach Penzance. About forty trains used the Cornwall line on the busiest Saturdays in the 1950s. At that time local trains from Plymouth to Truro and from Truro to Penzance supplemented the fifteen Monday-to-Friday Plymouth–Penzance services, most of them continuations of expresses from Paddington and the North.

Since then there have been several changes of policy. In the mid-1960s there were plans to reduce the Plymouth–Penzance service to a basic eight or nine trains each way daily, calling at all stations remaining open (apart from the halts within Plymouth), and to single most of the route. Lostwithiel, Par and St Erth were among stations planned for closure; the Newquay and St Ives branches would have used one of the present main-line tracks to make new junctions at Bodmin Road and Hayle.

The emphasis was heavily on economy rather than service, meaning that even the last Paddington express of the day that nominally served Cornwall terminated at Plymouth, passengers having to change into a multiple unit. Now, however, the county has the best express service in its history, there still being about fifteen Plymouth–Penzance trains and most of them still being through services, but faster and better balanced. The *Cornish Riviera Limited* (as it has been renamed) travels faster then ever and taking only 5 h. 23 m., arrives at Penzance (1972–3 service) at 4.53 p.m., only 23 minutes behind a very fast Leeds–Penzance *Cornishman.* The *Golden Hind* has been extended to and from Penzance; it departs in the morning at 5.10 a.m. and arrives back in the evening after allowing nearly six hours in London at 10.58 p.m. This would have been unthinkable in GW days.

But while the expresses now run like expresses through Cornwall and fewer impatient businessmen drive to Exeter to catch their train, or press for the introduction of an air service (or for that matter feel it necessary to patronize the nightly sleeper-only train at least in one direction), the priority given to long-distance passengers has been so total that local traffic patterns have sometimes been unnecessarily disturbed and connections with the surviving branch lines leave much to be desired. But maybe because the railway's total existence in Cornwall has so often publicly been said to be threatened, local management has worked hard, and reminders of 1930s excursion business have been occasional days of 60p excursions from Penzance and 40p from St Austell to Plymouth, up to 3,500 people being carried in a few hours.

Many fewer people use Truro's station; though awkwardly perched on a hill, in the 1950s it was used by 1,000 to 1,500 passengers daily excluding those just changing trains. It was the railway, of course, which confirmed Truro as virtually, if not technically, capital of Cornwall. It is a thriving market, shopping and administrative centre – though the population in 1971 was only 14,849. With a population of 42,029, Camborne–Redruth is the largest urban district in the county before the current local government reorganization, but in fact Camborne and Redruth each retain their individuality and their own stations: as has already been hinted, much local business has been lost to the roads but excursion traffic to Plymouth, particularly for football, still reaches considerable proportions. Before the war, 1,400 to 1,600 went from the two stations (the fare was only 5s) every time Plymouth Argyle was playing at home; the total from all stations averaged 3,000, and it has been stated that each time 200 men would buy suits from the Fifty Shilling Tailors in Plymouth. Indeed, Plymouth's influence as a shopping centre still pervades Cornwall, though today many Cornish towns have their own branches of multiple stores. Until 1961, the railway's share of the Plymouth traffic was maintained by the lack of a road bridge over the Tamar, and except at Saltash itself the opening of the bridge has not so far had as big an impact as might have been expected. Penzance, business and horticultural centre as well as resort, has gained enormously through its terminus status and is railhead to the Scilly Islands and the whole Land's End peninsula.

It was said that the main line was vital to Cornwall's economic survival; but the survival has been only partial. Indeed the county as a whole has lost ground, though the rapid rise of seaside places such as Newquay camouflages the worst of the inland depression. Tin partially replaced copper and there were minor booms in the first years of this century and in 1914–18. Then the market slumped. Special trains for emigrants left Camborne with heartbreaking farewells: in the town the street lights were extinguished as rates remained unpaid. The 1930s brought some recovery, but today only two mines survive, though mining engineering provides fair traffic. The unemployment rate is still high throughout West Cornwall. Fishing has dwindled almost as severely as mining, even since 1924 the number of Cornish fishermen having dropped from over 3,000 to little more than 500. Newlyn alone flourishes.

Today the holidaymaker reigns over Cornwall, but horticulture and the china-clay industry are both important and developed almost entirely round rail transport. By the late 1890s, for instance, up to 1,300 tons of new potatoes left West Cornwall annually, while in 1910 no fewer than 268 special trains (2,466 trucks) were needed for the broccoli traffic alone. In the 1950s, potatoes, broccoli and flowers, chiefly from Penzance, Marazion, St Erth, Gwinear Road, and the Helston branch, added up to 30,000 tons a year; much has been lost to road transport since, but business is still substantial at times, 1,000 trucks of broccoli – now officially known as winter cauliflower – leaving in the first half of January 1963 when roads were icebound. This, like the holiday traffic, is uneconomic to work, coming in sharp seasons and easily upset by the weather. The railway has also assisted in making Cornish agriculture healthier than ever before. In 1956, 6,310 rail tanks of milk (from St Erth, Camborne, Lostwithiel and Saltash) left by the main line, together with 24,027 cattle and 41,848 pigs. No cattle are now carried and milk is increasingly turned into dairy products within the South West. But fertilizers and other heavy agricultural requirements come in increasing loads by train. Petrol and clay are other heavy traffics, but increasingly the Cornwall main line is a passenger line and its contribution in knitting the county more closely to the rest of Britain from the passenger point of view is vital in solving the high unemployment rate. For businessmen will not take their works too far into the back of beyond, and even the retired, moving from other parts, like to think they can quickly reach London or Birmingham if the need presses.

Before leaving the main line, mention must be made of an accident between Doublebois and Bodmin Road on 16 April 1895 which paved the way throughout the country for the abolition of 0-4-4 tank engines for express trains. Actually, the two engines involved – they both left the rails at high speed – started as 0-4-2s and belonged to the 3521 class, built at Swindon in 1887. They had an unusually long rigid wheelbase. Owing to unsteady running, they had their rear pony trucks quickly replaced by more substantial bogies. But they still rolled violently. The accident occurred to an evening train from Plymouth, but evidence showed that the track had been damaged by the passage of the preceding *Cornishman*, also hauled by two of the same class.

The Board of Trade inspector, Major Anderson, shocked other companies which had so far used the type without accident by stating that 0-4-4 tanks were basically unsafe for expresses. But this view was given only as a recommendation, and the matter remained controversial until on 6 March 1898 an LSWR express hauled by such a machine plunged off an embankment near Tavistock in most spectacular fashion. Meanwhile some of the GWR's 3521 class had been relegated to branch duties, and shortly after this, on 31 October 1898, number 3542 plunged down an embankment when hauling the Falmouth mail. The driver was killed. Thereafter the remaining members of the class were converted into 4-4-0 tender engines, some remaining at work until the 1930s. On the other hand, the LSWR's 0-4-4 tanks, designed by locomotive engineer Dugald Drummond specially for West Country expresses, were retained unaltered, though taken to London for suburban work – on which, incidentally, one was involved in an accident as late as 1933.

It is worth adding that despite these and other accidents in Devon and Cornwall, there has to the best of my knowledge never been a passenger killed in either county as the result of a derailment or collision.

THE CHINA-CLAY DISTRICT AND NEWQUAY

If railway work is never uninteresting (this paragraph remains as written for the first edition in 1958; we shall see later that some facts have changed) a visit to the china-clay lines shows that it can be excessively difficult. The switchback gradients necessitate both banking assistance and the frequent pinning down of brakes by hand: the cloying slime rules out vacuum brakes. Hauls from the clay storage sheds to the harbours are unprofitably short, particularly as most trains have to be remarshalled at St Blazey, the hub of the clay system, since quays are limited and boats seldom arrive in advertised order. One day the clay may be piped in liquid form down to the harbours; meanwhile the railways still handle much of it. Indeed, a large proportion of it goes to America, and deep water vessels can be accommodated only at Fowey, where the railway quays – developed into their present form at the beginning of this century – have no road access. 600,000 tons annually leave Fowey alone. Altogether 800 special tip-end waggons and more than 20 locomotives are required.

China clay or kaolin (Kao Lin, the Chinese mountain where the substance was first discovered) is the decomposed felspar of granite. The deposits in Cornwall and at Lee Moor near Plymouth in Devon are among the largest in the world, and the best Cornish produce is unexcelled in quality. Hence the export trade. Clay is by now far the largest of Britain's raw material exports. The Cornish trade began as early as the mid-eighteenth century, but fortunately for several thousand metal miners who would otherwise have had to emigrate with the rest, the largest expansion was delayed just until many of central Cornwall's mines were closing. The clay has been used for more and more purposes, ranging from cosmetics to digestive tablets, but largely finds its way to paper manufacturers. Only a small proportion is sent to the potteries, where potter's or ball clay (as found in the Newton Abbot and Torrington districts of Devon and in Dorset) is in greater demand. China clay is white, and decomposed where it is now found. Ball clay is brown, and in the course of geological history was transported from its original situation. Returning to Cornwall; the white pyramids or tips on the skyline in the St Austell area and elsewhere are formed of quartz, a waste product which, like the other component of granite, mica, is separated from the clay by water action. Powerful hoses break down the clay rock into liquid form; the quartz is trapped and sent by miniature trucks to the top of the pyramids; the mica and clay continue together until the stream slackens and the mica, the heavier of the two, sinks first. The industry has become highly mechanized since the war and surprisingly little labour is now employed: there is often a marked contrast between the efficiency of the quarries and the inefficiency (if unavoidable) of the railways carting away the finished product.

The clay system was begun by J. T. Treffry, later chairman of the Cornwall Railway. In 1842 he opened a canal from Par to Ponts Mill and a standard-gauge railway from there to Bugle (Mollinis) by 1 in 10 Carmears Incline and the Treffry Viaduct, of ten magnificently proportioned arches across the Luxulyan Valley. Treffry decided to use horses – after hearing that the cost would be only three-quarters that of locomotives – except on the incline, where ropes were set in motion by a water-wheel. Later the line was extended beside the canal to Par harbour. In 1849 Treffry opened a similar railway from Newquay harbour through an inclined tunnel and over the Trenance Valley by a timber viaduct (rebuilt of stone in 1874) to St Dennis, with a branch to East Wheal Rose mine, Newlyn East. He regretted that the Trenance Viaduct could not be 'done for posterity' like Treffry Viaduct, but he was anxious to start the traffic and there had been delays owing to disputes with turnpike and other authorities: 'You have no idea how much my motives have been misconstrued by people viewing all that I do with a jealous feeling.' The Press, however, spoke kindly of the East Wheal Rose line's opening: 'This vast undertaking, of almost national importance, has been effected, we need scarcely observe, by the liberality of a single individual, that spirited capitalist, J. T. Treffry, Esq. of Place.'

Treffry died next year and his estate went into Chancery, where it remained for more than twenty years. By the time it was freed, a speculator, W. R. Roebuck, had appeared on the scene. He formed the Cornwall Minerals Railway, authorized on 21 July 1873 to take over and adapt for locomotives, or to build from scratch, the following lines, totalling 46¾ miles, all opened, after feverish activity, on 1 June 1874:

Fowey to Newquay, incorporating parts of Treffry's Par–Bugle and St Dennis–Newquay lines. Carmears Incline was replaced by a line in the Luxulyan Valley under Treffry Viaduct: part of the old route at the bottom is retained as a clay works siding, and the line from Luxulyan over the viaduct to a quarry was used until 1927. The viaduct remains intact. The St Blazey–Fowey section included stiff climbs on either side to 1,173-yd Pinnock Tunnel;

Bugle to Carbus, more recently called Carbis, and St Dennis Junction (at first called Bodmin Road) to Melangoose Mill (later extended to Meledor Mill and known as the Retew branch), both still actively used for clay traffic (a later branch off the Carbis one to Wheal Rose closed in the 1950s);

Treloggan (now Tolcarn) Junction (outside Newquay) to East Wheal Rose, extended first to Treamble and later to Gravel Hill. Gravel Hill extension was taken up in 1888. Part of the branch was incorporated in the Perranporth–Newquay passenger line. The rest was closed in 1917, and reopened in 1926, but quickly fell into disuse again and though never officially reclosed was taken up after the war;

The Newquay & Cornwall Junction Railway, opened on 1 July 1869 from the Cornwall Railway at Burngullow to Drinnick Mill, the only broad-gauge section of the group, and Drinnick Mill–St Dennis Junction. The Act stipulated that broad as well as narrow rails were to be provided on the latter. When the Cornwall Railway pressed the point with Chancery proceedings they

Cornwall and West Cornwall Railways.

UP TRAINS.	WEEK DAYS.		C			D	SUNDAYS. E	
		1 2 3 A. M.	N.M. 1 & 2 A. M.	1 2 A. M.	mail. 1 & 2 P. M.	1 2 3 P. M.	1 2 3 A. M.	mail. 1 2 3 P. M.
enzance... dep.		6 25	10 0	11 15	3 50	6 30	6 50	3 50
Marazion Road.		6 31	10 6	11 21	3 56	6 36	6 55	3 56
t. Ives Road ...		6 40	10 15	11 31	4 5	6 46	7 4	4 5
Hayle............		6 47	10 22	11 38	4 12	6 53	7 10	4 12
winear Road ..		6 57	10 32	11 48	4 22	7 3	7 22	4 22
amborne		7 5	10 40	11 58	4 30	7 12	7 31	4 30
ool		7 11	10 46	12 3	4 36	7 18	7 37	4 36
edruth		7 18	10 53	12 12	4 43	7 25	7 43	4 43
corrier Gate ...		7 25	11 0	12 20	4 50	7 32	7 49	4 50
hacewater	F	7 31	11 6	12 26	4 56	7 38	7 54	4 56
ruro......... arr.	1 2 3	7 43	11 18	12 40	5 8	7 55	8 5	5 8
	A. M.							
Falmouth dep.	6 5	7 15	10 45	...	4 35	7 20	7 30	4 35
Penryn	6 15	7 25	10 55	...	4 45	7 30	7 40	4 45
Perranwell	6 25	7 36	11 5	...	4 55	7 40	7 50	4 55
Truro...... arr.	6 35	7 45	11 15	1 & 2	5 5	7 50	8 0	5 5
				P. M.				
ruro .... dep.		7 52	11 28	12 50	5 15	8 2	8 12	5 15
ramp'd-rd arr.		8 9	11 45	1 9	5 32	8 19	8 29	5 32
urngullow „		...	11 58	...	...	8 31	8 41	...
t. Austell.. „	—	8 24	12 4	1 27	5 47	8 36	8 46	5 47
ar .... „		8 34	12 14	1 39	5 57	8 46	8 56	5 57
ostwithiel. „		8 47	12 27	1 55	6 9	8 58	9 8	6 9
odmin-Rd. „	1 2 3	8 56	12 36	2 7	6 18	9 7	9 17	6 18
oublebois. „	A. M.	9 13	12 53	...	6 35	9 24	9 34	6 35
iskeard ... „	7 50	9 23	1 4	2 39	6 45	9 34	9 44	6 45
enheniot.. „	7 59	9 32	1 13	...	6 54	9 43	9 53	6 54
t. Germans „	8 12	9 45	1 27	2 59	7 7	9 56	10 6	7 7
altash ...... „	8 26	9 59	1 41	3 13	7 19	10 9	10 18	7 19
evonport.. „	8 35	10 8	1 51	3 22	7 27	10 18	10 28	7 27
lymouth ... „	8 41	10 15	2 0	3 30	7 35	10 25	10 35	7 35
		1 & 2 A. M.					1 & 2 P. M.	mail. 1 & 2
lymouth... dep.	9 5	10 25	2 15	4 10	7 45	...	...	7 45
xeter ...... „	11 45	12 25	4 45	7 30	10 12	...	3 30	10 12
ristol ...... „	3 50	2 45	7 0	...	12 50	...	6 30	12 50
addington arr.	9 25	6 0	10 20	...	4 35	...	11 10	4 35

C—Third lass from Falmouth, Penzance, and intermediate tations, to St. Germans inclusive, to Bristol and Exeter stations bove Exeter, and to certain Great Western stations; also third lass from West Cornwall and Cornwall stations, including Devonport, and from South Devon stations to Paddington. Third class Wednesdays and Saturdays only, from West Cornwall stations and from Falmouth, Penryn, and Perranwell, to Truro; also third class from West Cornwall stations to Redruth, Fridays.

D—Third class from West Cornwall stations to Bristol and Paddington. Such passengers must leave Plymouth at 6.35 a.m. he following day.

E—Third class from Cornwall and West Cornwall stations to ertain Bristol and Exeter and Great Western stations.

F—This train also runs on Sundays, and returns from Truro or Falmouth at 8.15 a.m., arriving at Falmouth at 8.45 a.m.

DOWN TRAINS.	WEEK DAYS. G			H	I		SUNDAYS. G	
	mail. 1 2 P.M.	1 2 3 A.M.	N.M. 1 & 2 A.M.	1 & 2 A.M.	Exp. 1 & 2 A.M.	Exp. 1 & 2 A.M.	mail 1 & 2 P.M.	N.M. 1 2 3 A.M.
Paddington. dep.	8 10	...	...	6 0	9 0	11 45	8 10	...
Bristol	12 30	...	6 15	10 30	12 30	2 26	12 30	6 30
Exeter	3 0	...	8 40	1 40	3 20	4 10	3 0	9 45
Plymouth ...arr.	4 40	...	10 48	4 21	5 20	6 0	4 40	12 26
	A.M.						1 2 3	P.M.
Plymouth ...dep.	4 50	6 55	11 0	2 30	6 10	8 0	4 50	2 30
Devonport...arr.	...	7 1	11 6	2 35	6 15	8 5	...	2 35
Saltash ...... „	5 1	7 9	11 14	2 46	6 25	8 15	5 1	2 45
St. Germans „	...	7 23	11 28	2 59	6 39	8 29	...	2 59
Menheniot ... „	...	7 37	11 42	3 13	6 54	8 43	...	3 13
Liskeard...... „	5 30	7 47	11 52	3 23	7 4	8 55	5 30	3 23
Doublebois.. „	...	7 57	12 2	3 33	7 14	1st, 2nd, and 3rd class.	...	3 33
Bodmin-Rd.. „	5 49	8 12	12 17	3 48	7 29		5 49	3 48
Lostwithiel.. „	...	8 22	12 27	3 57	7 38		...	3 57
Par .... „	...	8 34	12 39	4 10	7 50		...	4 10
St. Austell .. „	6 15	8 47	12 52	4 24	8 3		6 15	4 24
Burngullow. „	...	8 52	12 57	4 30	8 8		...	4 30
Gramp'nd-rd „	...	9 4	1 9	4 42	8 19		...	4 42
Truro ....arr.	6 45	9 21	1 26	5 3	8 36		6 45	5 3
Truro ......dep.	6 50	9 30	1 35	5 20	8 45		6 50	5 20
Perranwell ar.	...	9 40	1 45	5 30	8 55		...	5 30
Penryn	7 5	9 50	1 55	5 40	9 5		7 5	5 40
Falmouth	7 15	10 0	2 5	5 50	9 15		7 15	5 50
Truro......... dep.	6 53	9 27	1 32	5 10	8 43		6 53	5 10
Chacewater	7 5	9 39	1 48	5 23	8 58		7 5	5 23
Scorrier Gate ...	7 10	9 44	1 54	5 28	9 4		7 11	5 28
Redruth	7 18	9 52	2 2	5 36	9 12		7 18	5 36
Pool	7 24	9 58	2 8	5 44	9 18		7 24	5 44
Camborne	7 31	10 5	2 15	5 51	9 25		7 31	5 51
Gwinear Road..	7 39	10 13	2 23	5 59	9 33		7 39	5 59
Hayle............	7 48	10 22	2 32	6 8	9 41		7 48	6 8
St. Ives Road...	7 54	10 28	2 38	6 14	9 47		7 54	6 14
Marazion Road.	8 4	10 38	2 47	6 23	9 57		8 4	6 23
Penzance ...arr.	8 10	10 44	2 53	6 28	10 4		8 10	6 28

G—Third class from Paddington and certain Great Western stations to South Devon, Cornwall, and West Cornwall stations where the Mail calls.

H—Third class from Great Western stations (except Bristol) to South Devon (except from Paddington to Exeter), Cornwall, and West Cornwall stations.

I—Third class from Paddington only, to Exeter, and to South Devon, Cornwall, and West Cornwall Stations.

N.B.—The Express Limited Mail Train leaves Paddington at 9 p.m., carrying a limited number of first class passengers booked from London only. This overtakes the mail train at Bristol.

☞ An express train leaves Paddington at 11.45 a.m., reaching Exeter at 4 p.m., and Plymouth at 6 p.m., in time for the 6.10 p.m. train from Plymouth for Falmouth and West Cornwall.

⁂ Third-class return tickets (one shilling each) are issued on Mondays and Fridays by the trains leaving Truro for Falmouth at 1.35 a.m. and 5.20 p.m.; also by the trains leaving Falmouth for Truro at 10.45 a.m. and 4.35 p.m.

Fowey and Newquay.

DOWN TRAINS.	WEEK DAYS ONLY.									
	P.M.	A.M.	A.M.	A.M.	A.M.	A.M.	P.M.	A.M.	A.M.	P.M.
LONDON (Padd.) dep.	9 0	......	......	......	......	5 30	......	10 15	11 45	......
Plymouth (Millbay) „		6 50	—	9 35	11 15	1 28	—	4E0	6 20	—
Par ....dep.		8 40	10 5	11 16	12 55	3 18	4 25	5 35	7 50	8 40
NEWQUAY — „	—	—	8 55	—	11 30	2 0	—	—	5 20	7 20
St. Blazey .... „	......	8 46	10 11	11 23	1 2	3 24	4 31	5 41	7 58	8 46
FOWEY — — arr.	—	8 58	10 22	11 34	1 14	3 36	4 43	5 52	8 8	8 58
Par ....dep.	......	8 40	......	11 16	12 55	......	North Rd.	5 35	7 50	......
St. Blazey — — „	—	8 45	—	11 22	1 2	—		5 40	7 56	—
Bridges (for Luxulyan) „	......	8 56	......	11 32	1 12	......		......	8 6	......
Bugle — — — „	—	9 2	—	11 39	1 18	—		5 57	8 12	—
Victoria (for Roche) „	......	9 9	......	11 46	1 25	......		6 5	8 21	......
St. Columb Road — „	—	9 23	—	11 59	1 38	—		6 20	8 35	—
NEWQUAY ......arr.	......	9 36	......	12 12	1 51	......		6 33	8 48	......

UP TRAINS.	WEEK DAYS ONLY.									
	A.M.	A.M.	A.M.	P.M.	P.M.	P.M.	P.M.	P.M.	P.M.	
NEWQUAY.......dep.	......	8 55	11 30	2 0	......	......	......	5 20	7 20	......
St. Columb Road „	—	9 11	11 46	2 16	—	—	—	5 36	7 35	—
Victoria (for Roche) „	......	9 25	12 3	2 31	......	......	......	5 50	7 50	......
Bugle — — — „	—	9 32	12 10	2 39	—	—	—	5 59	7 58	—
Bridges (for Luxulyan),	......	9 38	12 16	2 44	......	......	......	6 5	8 7	......
St. Blazey — — arr.	—	9 48	12 25	2 54	—	—	—	6 14	8 16	—
Par .... „	......	9 54	12 32	3 0	......	......	......	6 20	8 22	......
FOWEY — dep.	8 0	9 30	12 10	2 40	3 45	4 50	5 22	6 0	—	8 13
St. Blazey .... „	8 18	9 52	12 30	2 58	4 3	5 7	5 40	6 18	......	8 27
NEWQUAY arr.	9 36	—	1 51	—	—	—	6 33	8 48	—	—
Par .... „	8 20	9 54	12 32	3 0	4 5	5 9		6 20	8 22	8 29
Plymouth (Millbay) arr.	10 10	11 30	1E59	5 55	5 55	—	—	8E4	10 15	10 15
LONDON (Padd.) ..	4 30	6 30	7 45			......	......	4 0		

Falmouth Branch.

DOWN TRAINS.	WEEK DAYS.																	SUNDAYS.						
	A.M.		1,2,P. A.M.		P.M.		P.M.		P.M.		P.M.		P.M.	P.M.		1,2,P. P.M.		A.M.	1,2,P. A.M.		1,2,P. P.M.			
TRURO ....dep.	7 3	......	9 28	......	12 10	......	1 48	......	4 10	......	5 45	......	6 18	7 55	......	8 32	......	7 3	9 50	......	5 45	......	......	......
Perranwell — — — „	7 13	—	9 38	—	12 20	—	1 58	—	4 20	—	5 55	—	6 28	8 5	—	8 45	—	7 13	10 0	—	5 55	—	—	—
Penryn .... „	7 22	......	9 48	......	12 30	......	2 8	......	4 30	......	6 5	......	6 38	8 15	......	8 55	......	7 22	10 10	......	6 5	......	......	......
FALMOUTH — — arr.	7 32	—	9 58	—	12 40	—	2 18	—	4 40	—	6 18	—	6 48	8 25	—	9 5	—	7 32	10 20	—	6 15	—	—	—

UP TRAINS.	WEEK DAYS.																	SUNDAYS.						
	1,2,P. A.M.			1,2,P. A.M.		A.M.		A.M.		P.M.		P.M.	P.M.		1,2,P. P.M.	P.M.			1,2,P. A.M.		1,2,P. P.M.		P.M.	
FALMOUTH ....dep.	7 12	......	......	8 35	......	10 38	......	11 2?	......	2 40	......	3 20	5 0	......	7 5	8 45	......	......	9 0	......	2 50	......	5 0	......
Penryn — — — — „	7 23	—	—	8 45	—	10 48	—	11 [illegible]0	—	2 50	—	3 30	5 10	—	7 15	8 56	—	—	9 10	—	3 0	—	5 10	—
Perranwell .... „	7 32	......	......	8 55	......	10 58	......	11 40	......	3 0	......	3 39	5 20	......	7 25	9 5	......	......	9 20	......	3 10	......	5 20	......
TRURO — — — — „	7 42	—	—	9 5	—	11 [illegible]	—	11 50	—	3 10	—	3 50	5 30	—	7 35	9 15	—	—	9 30	—	3 20	—	5 30	—

Cornwall Railway timetable for summer 1876.
Newquay and Falmouth line timetables for summer 1893.

were laid – but not in such a manner that they could be used. So until 1892 break of gauge remained at Drinnick Mill, where clay and coal traffic has always been heavy. Between 1909 and 1922 the Carpella Break prevented through Burngullow–St Dennis communication: as customary, the vendors of the land used by the railway retained mineral rights, and after proceedings ending in the House of Lords it was ruled that china clay was a mineral and that the Carpella United Clay Company had the right to require the removal of the track while excavating. Drinnick Mill boasted its own signal box for many years after the gap was restored, but in 1965 the route was closed from St Dennis Junction to Pankandillack Siding, leaving $4\frac{1}{2}$ miles of track up from Burngullow.

Clay quarrying was still in its infancy in Roebuck's day, and when a Cornish Consolidated Iron Mines Corporation failed in a guarantee of traffic, the Minerals Railway desperately tried to augment income by a Fowey–Newquay passenger service, started on 20 June 1876. Next year the line was leased to the GWR (the Minerals Company remained in existence till 1896, receiving steadily increasing rental as the result of better clay traffic), and on 1 January 1870 a double-line curve was opened from St Blazey to Par. The Minerals line passenger trains from both Newquay and Fowey – the Fowey ones had to reverse at St Blazey – now pulled up alongside the Truro–Plymouth ones at Par, but different gauges prevented through working until 1892.

On 1 June 1869 a broad-gauge Lostwithiel & Fowey Railway had opened for goods to Carne Point, a mile short of Fowey. For some years it carried clay, but at so low a price in opposition to the Minerals line (a ton between Burngullow and Fowey *via* Lostwithiel eventually cost only 2s 2d) that it ran out of funds and when bridges became unsafe it had to close – on 1 January 1880. It was reopened (as a narrow line, of course) to join the Minerals at Fowey proper on 16 September 1895. For a time Fowey thus had two passenger routes, but the St Blazey one, through Pinnock tunnel, was closed to ordinary passengers on 8 July 1929 and to workmen from 31 December 1934. The Lostwithiel trains ceased for part of both world wars but lasted until 4 January 1965. Fowey still owes much of its prosperity to the clay harbour business. Both lines remained open until August 1968, when as the result of a pact between English China Clays Ltd and British Railways the whole method of handling the clay traffic was changed.

English China Clays had long wanted road access to Fowey, especially since Par Harbour, shallow at the best of times and waterless at some low tides, was at capacity. Fowey in any case was the only harbour capable of taking trans-Atlantic ships. British Railways therefore sold the trackbed of the St Blazey–Fowey route through Pinnock Tunnel, which has now been converted into a road, subject to guarantees of traffic tonnages from the company. Initially the guarantee was that a minimum of 500,000 tons of clay annually would reach Fowey by rail, but this has since been increased to 600,000 and could rise further yet as the result of new plans between the railways and the company. Since this book first appeared, the total clay business has continued to increase dramatically, over 1,000,000 tons annually leaving Fowey alone. After various experiments and teething troubles, the fleet of china-clay tip-waggons, now numbering nearly 900, has been fitted with vacuum brakes, despite the forecast that this would be impossible. St Blazey remains the hub of the system, but apart from small amounts still going into Par Harbour, the export clay is now all routed – in about seven daily trains – *via* Lostwithiel. At Lostwithiel trains reverse for Carne Point reception sidings, short of Fowey proper. By the time this edition reaches the bookshops, the clay waggons will even be covered for the first time in history; for generations the clay companies complained that the railways left the clay exposed to the elements.

Part of the agreement between English China Clays and the railways is that more clay for destinations in other parts of Britain goes by train. A daily clay liner to the Potteries has to be duplicated on occasions, and there is now even a weekly train fitted with airbrakes leaving St Blazey at 10.10 one morning and arriving at Glasgow at 6.15 the next.

But even better could be to come, since the environmental harm caused by quarrying for building sand in the Home Counties and by piling waste sand into the grey-white dunes that dominate the clay field could be mutually cancelled by the railways carrying the stuff from Cornwall to London. Even ten years ago such a plan might have been dismissed as a pious hope, but the environmental lobby is now powerful.

These other clay lines have to be mentioned: Bugle to Goonbarrow, including a 341-yd tunnel, opened in 1893; it survives as far as New Cauldown Siding. The steeply-graded remainder of the branch including a reversal at a dead end was always particularly difficult to work. The Trenance Valley line, opened as far as Bojea sidings on 1 May 1920 and on to Lansalson on 24 May the same year was never as busy as most of the clay routes and closed in 1966.

A 3 ft 1 in tramway from Gothers to the Burngullow–St Dennis Junction line closed in 1931 after a short career. Finally, but not least in historical importance, the 2 ft 6 in. gauge Pentewan Railway, an isolated, individualistic concern which played a surprisingly energetic role in the development of the clay industry, at first concentrated immediately round St Austell. Opened in 1830 (three years later it carried 3,296 tons of clay at 4d a ton), and worked by a combination of gravitation and horses, the four-mile line connected the town with Pentewan Quay. Some tin was taken too, and coal formed a return load. Some time after an Act of 1874, though not immediately, owing to opposition from a landowner, the tramway became a fully-fledged locomotive line. Closure came in 1916.

While the Fowey end of the Cornwall Minerals Railway was always dominated by mineral business, and passenger services at 'Troy Town' had a chequered history, the reverse is the case at the other end, at Newquay, where the one-mile mineral line from the station to the harbour was used for the last time in 1926 but summer week-end passenger traffic once reached very large proportions. A fishing village a century ago, and having only 1,000 people in 1871, today the resort is by far Cornwall's most important with a 1971 population of 15,017, while literally thousands camp in the surrounding area. For many years, peak-season visitors overflowed into the villages of the china clay area, a few of which have indeed set up in tourism on their own account. The train service from Par to Newquay increased with the trade, and during the 1920s and 1930s various improvements were made, including the extension of Newquay station to accommodate 15-coach trains and the doubling of one section across Goss Moor. 'Castles' could use the branch; but it was built as a mineral line and has always been unsatisfactory. In steam days often three engines were used to take heavy trains up Luxulyan Bank, with its stretches of 1 in 37. In the 1950s up to 6,000 visitors from beyond Bristol arrived on a peak Saturday, but the slowness of the journey encouraged air as well as road competition. Newquay remains the only Cornish branch-line town still to receive through trains from London and the Midlands and North on summer Saturdays, but though still well patronized these are both less numerous and shorter than used to be the case. Sections between crossing loops on the single track have been lengthened, and Newquay's station buried behind a redevelopment scheme. The ordinary all-the-year-round service of six trains is weak, not one train connecting into an up London express. Par is now the junction point again, the experiment of running the branch trains on to Bodmin Road being abandoned.

OTHER BRANCH AND MINOR LINES

The TREGANTLE MILITARY Railway started at Wacker on the upper reaches of the Lynher River (a branch of the Tamar) and served Scraesdon and Tregantle Forts designed to protect the sea approaches to Plymouth. Opened about 1893 and of standard gauge, it had a career of only about ten years, probably because of the high costs of operation, which included perpetually dredging a channel for barges to reach the quay at Wacker, a cable-incline worked by a stationary steam engine, two locomotives (one below and one above the incline), and frequently hand-shunting by working parties of the trucks on an inadequate layout – especially at Wacker. The trucks, and possibly also the track, were originally destined to help in the Sudanese campaign and were actually shipped but brought home and sent to Cornwall as the result of the assassination of General Gordon and the fall of Khartoum. Thus it is that today's industrial archaeologist can trace the route of one of the West's least-known railways and see the rusting engine shed and the pier slipping into the river at Wacker in one of the quietest spots in Cornwall, albeit so close to Plymouth as the crow flies.

The Liskeard & CARADON Railway was opened from

Moorswater (in a valley below Liskeard where it joined the Liskeard & Looe Union Canal) to South Caradon in November 1844 and to Cheesewring quarries in March 1846. Trucks loaded with tin ore and granite descended the standard-gauge line by gravity, and were returned empty by horses. When traffic reached 40,000 tons annually, the Canal built a continuation of the railway on its banks to LOOE. Opened on 27 December 1860, from 1862 it was worked by the Caradon Railway, which also began using locomotives on its own system and added a branch from South Caradon to Tokenbury Corner. In the 1880s this was extended round Caradon Hill back to the original Cheesewring line, of which the incline section between South Caradon and the new junction was closed. Passengers were carried on the Looe line from 11 September 1879; passenger accommodation was provided on the Caradon one from 1860, but before then passengers could travel in the mineral waggons by obtaining a 'free pass' if they paid for a parcel, umbrella, etc. It was only on 15 May 1901 that the sharply graded loop from Coombe Junction to the separate branch terminus at right angles to Liskeard's main station was opened. Passenger traffic immediately trebled, but the mines' heyday had passed. The whole Caradon system, 12 miles including the Kilmar granite tramway, was abandoned on 31 December 1916, and today the Looe quay where boats once loaded the ores and granite – the track continued beyond the station – is a car park. Visitors who still go by train find the journey slow, involving a reversal at Coombe Junction, but infinitely more fascinating than it would be had a 1935 scheme for a direct line from St Germans matured. The GWR worked both lines from 1909, absorbing the Caradon Company the same year and the Looe one in 1923. Clay traffic is still carried from Moorswater.

Although not built till many years after a Padstow–Fowey scheme had been mooted by William James, rival to George Stephenson as 'Father of Railways', the BODMIN & WADEBRIDGE was Cornwall's first locomotive line. With a mileage of 22, its main line ran from Wadebridge to Wenford Bridge. It was opened to Dunmere on 4 July 1834 (when the Bodmin branch was also opened) and throughout on 30 September (when traffic began on the Ruthern Bridge branch). Including two engines and 40 waggons, the cost, borne by local people, was only

Bodmin and Wadebridge Railway.

No. 279 Dec: 30th 1873

Mr. Tippett

Please to supply for the use of the BODMIN AND WADEBRIDGE RAILWAY:

1 Pint Cocoa to
Wm Jenkinnick
Jos. Gregg

H. Kyd
Supt

about £35,000. Sea sand for manure and minerals for export (sizeable boats then reached Wadebridge) were the chief loads. But several times a week a whistle told Wadebridge folk that a passenger train was about to leave for Bodmin – where coal fell from 25s to 15s a ton. The first 'Day Book', kept by the superintendent, H. Kyd – who constantly authorized his staff pints of cocoa on the Company! – is a catalogue of mishaps, such as 'Broke off Chimney of the Engine' and 'Stopped the Engine in consequence of the wheels being of different sizes'. The weather's influence was profound. Snow made it 'unprofitable to work', and during the thaw: 'When at the Bridge under Pendavey, the weather being very rough and the evening very dark, found it necessary to return.' Spring brought a standstill, 'fearing the consequences of the dry weather'. But Kyd was a good businessman and on 13 April 1840 a special carried the greater part of Wadebridge to see an execution at Bodmin.

Bodmin Road to BODMIN, always narrow gauge, began its career as a main line feeder on 27 May 1887, and on 3 September 1888 was extended round the town to Boscarne Junction on the Bodmin & Wadebridge, whose section on to Wadebridge was improved (the granite sleepers typical of early Cornish lines being replaced) to accommodate GWR trains. As seen in Chapter V, Wadebridge was reached by the LSWR's North Cornwall line on 1 June 1895. The North Cornwall's final section, beside the windswept, sandy Camel to PADSTOW, followed on 27 March 1899, after which the district's passenger arrangements changed little for 60 years, shuttle services from Wadebridge running to both Bodmin North (the former Bodmin & Wadebridge station) and Bodmin General (formerly GWR). This was justified more by poor bus services than by activity at Bodmin, which has lost much to Truro. Ruthern Bridge was closed on 30 December 1933, but china clay has renewed life for the Wenford Bridge line, until 1964 still visited daily by a 2-4-0 Beattie tank built in 1874 (see frontispiece). Padstow remained a busy port even between the wars, when the Southern built a new fish station. Coal, empty boxes and ice for the ships – not mainly local ones – as well as 1,000 waggonfuls of fish in a spring season gave the station an atmosphere of considerable bustle. Wadebridge relied heavily on the passenger service, totalling fifty trains a day to Padstow, up the North Cornwall line to Holwill Junction and Bodmin North on the former LSWR routes and to Bodmin and Bodmin Road on the former GWR. Animosity continued after nationalization, Western trains not being allowed to serve the halts between Bodmin and Wadebridge. Eventually a reduced Bodmin Road–Padstow service was introduced, passengers from Bodmin North using a railbus to a new Boscarne Junction Exchange Platform, though a three-coach unadvertised school train was also used. The whole system closed to passengers from 30 January 1967. Padstow had already lost its freight service but freight continues to Wadebridge and clay still leaves Wenford Bridge by rail. Wadebridge Quay was however closed in 1973. Today the small town is almost entirely a resort.

After the West Cornwall's extension to the Cornwall's Truro station, Penwithers Junction to NEWHAM became a goods branch, except that one daily passenger train used it until September 1863. Freight, including coal for the gasworks, remained heavy until the 1950s but the line was closed on 6 November 1971. The final section of the Cornwall Railway, from Truro to FALMOUTH, was opened on 24 August 1863, a broad-gauge single line running parallel to the West Cornwall's narrow one as far as Penwithers Junction, where it cut diagonally across the Newham branch. 'An extensive system of signal machinery', believed to be the first interlocking in the West, was installed; but later access to Newham was from the Falmouth line. The branch's engineering works included eight of Brunel's timber viaducts, the last, Collegewood, 318 yds and 100 ft high, remaining in use until 21 July 1934. Originally the Falmouth terminus was to have been at Greenbank, but the present site was chosen to enable a convenient connection (opened in January 1864) with the new docks. Though Falmouth never recovered as a general port and expresses quickly began deserting it for Penzance, specialist ship repairing became important. When the GWR introduced a new daily late afternoon express from Paddington after the 1939–45 war, its original destination was Falmouth, and into the 1960s through trains carried many of the 1,500 holiday makers arriving by train at Falmouth on summer Saturdays. The local service had 15–21 weekday trains for many decades, and this was the only GWR branch to enjoy an all-the-year-round Sunday service. With no signalling beyond Penwithers Junction, it is now possible to run about a dozen trains even on busiest days, but these carry a remnant of commuting business as well as people making main-line connections. On 7 December 1970 the original terminus, by then much run down, was replaced by a single platform beside a new dead-end, shortening the branch by about a quarter of a mile.

In contrast, the Chacewater–Newquay line has few engineering works. It takes an indirect course and is remarkable for its range of scenery, including sea views at PERRANPORTH, a modern resort hampered by the lack of an old-established nucleus. Chacewater to Perranporth was opened on 6 July 1903 and the rest (the East Wheal Rose mineral branch was used between Shepherds and Tolcarn Junction) on 2 January 1905. At first the branch met the main line by a triangular junction, but in 1926 as an economy measure the three signal boxes were closed and the single line continued beside the main line to Chacewater station. This was a fairly busy branch, with 250,000 passengers annually in the 1950s; but it was closed from 4 February 1963.

Cornwall's most important purely mineral line was authorized on 17 June 1824 (£19,500 of the estimated £22,500 had already been raised) and opened on 30 January 1826. It was the REDRUTH & CHASEWATER [*sic*], of 4 ft gauge, whose nine-mile main line served some of the richest copper mines and ran from near Redruth *via* Lanner, the summit, to Point Quay on Restronguet Creek on the Fal River. A branch from Lanner to Wheal Beacham soon followed, and in 1853 one to Chacewater was started, but abandoned after the collapse of the mines it was to serve. Traffic averaged 60,000 tons annually, including considerable quantities of South Wales coal brought by the sailing ships which came to collect the ore. In 1854 the railway began using coal itself, having acquired two 0-4-2 side tank engines, cab-less and with tall chimneys, *Smelter* and *Miner*. Traffic increased (91,000 tons in 1861) until the collapse of copper; then receipts dwindled away and in 1879 a receiver had to be appointed. But it was not until the Company had been in Chancery thirty-six years that the closure of the line was enforced – on 25 September 1915. Few mines remained by then. This railway, typical of so many in the West, played a valuable part in developing the industry it was born to serve and had no hope of survival when that industry was killed by external circumstances.

The first such line was the POLDICE Tramway, which linked mines with Portreath's entirely artificial harbour on the north coast. Its first rail was laid by Lord de Dunstanville on 29 October 1809. The cost was £20,000. When the line opened, probably in 1810, it was reported: 'The wheels of the carriages run on cast iron, which facilitates in an extraordinary manner the progress of the vehicles and greatly lessens the force of animal exertion.' But the typical L-shaped plates were in short lengths and the road by no means smooth. Little has been discovered about the working of this line, though what is believed to be the carriage – a four-wheeled open affair built of yellow pine – originally used by the directors and officials has been preserved. Poldice, once mighty among copper mines, was exhausted before most, in 1856 being virtually at a standstill. Other mines were also served by the tramway, but this seems to have gone out of frequent use about 1865, though not dismantled until this century. It is difficult for us to recapture the hopes and fears these ventures roused. Of an earlier unsuccessful scheme to link Portreath with Dolcoath mine near Camborne, on the one hand we read 'The Carriers are all frightened in this neighbourhood with the sound of Cast Iron Rail Roads about to be fixed from Portreath . . . for the carriage of ores and Coals' and on the other 'They [the shareholders – Lord de Dunstanville held just over half the shares] calculate the probable profits of this concern at £15 per cent at least.'

The PORTREATH branch of the Hayle Railway, opened on 23 December 1837, carried so much ore and coal that in 1846 it was necessary to add the inner basin to the harbour. Often there would be 20 vessels loading or discharging at the same time, and waggons were continually being shunted around the extensive siding system or pulled up the long rope-worked incline to the cliffs above. Even at the beginning of this century up to 200 boats a year called, but regular railway traffic to the harbour ceased about 1930; the rest of the branch, from the junction with the main line to North Pool Siding, closed on 1 April 1938. The other Hayle Railway line, to TRESAVEAN, closed in 1932; it crossed the Redruth & Chasewater Railway on the level, but there was no exchange of traffic. The last tin area goods branch to close in the 1960s was that from near Camborne to NORTH ROSKEAR, the siding into Holmans works being the only piece of rail left working in the Camborne area other than the main line.

As it wound inconspicuously across the undulating plateau, the HELSTON branch seemed almost ashamed of itself. Opened by a local Company on 9 May 1887, it had an unexciting career and is perhaps best known for the fact that the first GWR bus service ran in connection with it. The 22-seaters began plying from Helston to the Lizard on 17 August 1903 and saved the expense of £85,000 on a Light Railway, for which an Order had been obtained. Though the vehicles soon had to be withdrawn temporarily owing to damage to the roads, numerous other railway bus routes followed and by 1927 the GWR's buses alone were carrying 8,000,000 passengers annually. (In 1929 the Western National Company took over the West Country routes.) Incidentally, the success of the pioneer buses saved in the nick of time the building of two other Light Railways for which Orders had also been obtained – the Padstow, Bedruthan & Mawgan, and the Penzance, Newlyn & West Cornwall. Helston combines business as a market, tourist and Naval airfield centre, and goods traffic, including broccoli and flowers, was heavy until complete closure from 5 October 1964. Passenger trains were withdrawn on 3 November 1962 and all traffic ceased in 1967.

When the Hayle Railway was reconstituted as the West Cornwall, and was taken above Hayle by a viaduct, a waggon lift down to the original station and siding system was intended. However, the steeply graded HAYLE WHARVES branch, equipped with what may have been Britain's first sand drag, was built instead. Broad-gauge rails were added in October 1877, when the line was also extended by a drawbridge along the quays. Traffic, once very heavy, is now slight and many of the sidings have been removed. Hayle Harbour's chief trade is in coal, used on the spot by the power station. The Harbour branch closed to ordinary traffic in 1967 though some private siding business remained.

A journey along the cliffs on the ST IVES branch on a fine day, when the sea, nearer Mediterranean blue than any other British waters, contrasts brilliantly with the white sand, is unforgettable. Holidaymakers who arrive by road miss much of the thrill. On summer Saturdays long-distance rail arrivals total up to 1,500, while almost everyone staying elsewhere in West Cornwall visits St Ives, for though today entirely a resort its atmosphere is genuinely Cornish, dating back to mining-fishing days. Opened on 1 June 1877, the branch was the last to be built broad gauge. In 1888 a narrow rail was added as far as Lelant Quay, but soon the quay's sidings were abandoned. Engineering works are heavy: in 1955 the branch was temporarily closed for the building of Porthminster Viaduct, just outside St Ives. Goods trains were withdrawn in 1964, and the layout at St Ives reduced to a simple dead-end without points or signals as part of the short-lived Plymouth Division's campaign to save a number of Cornish lines from closure.

Finally, the most westerly railway in England, the PENLEE QUARRY line. The quarry is at Newlyn, and the 2-ft gauge system was usually active only when a ship was loading at the harbour, but then up to ten trains each carrying 40 tons ran hourly. Every few minutes a full train was 'crossed' by an empty one on the loop mid-way down the main line, and the rattle of the rake of trucks being pulled or propelled by the small diesels over the worn rails and points scarcely ever died away. The diesels took over in 1940 from steam locomotives, one of which is illustrated as plate 24. In August 1972 the line was replaced by a conveyor belt.

CHAPTER VIII

South West Wiltshire and East Somerset

THE WILTS, SOMERSET & WEYMOUTH SYSTEM

Despite early ambitious plans South West Wiltshire, then a district of large sheep farms and small towns highly dependent on woollen manufacturing, was not assured of its place on the railway map until the beginning of the gauge war between the Great Western and the London & South Western Railways. In 1844 the LSWR proposed a line from Basingstoke to Swindon, in the heart of GWR territory. The GWR retaliated by promoting two nominally independent companies to extend the broad gauge, the Berks & Hants, and the Wilts, Somerset & Weymouth, both of which obtained their Acts on 30 June 1845.

The Wilts, Somerset & Weymouth was to leave the main GWR at Thingley, two miles west of Chippenham, and go to Salisbury, with a line branching off at Upton Scudamore for Yeovil and Weymouth. Branches were to run to Bradford, Devizes, Radstock and Sherborne. It was claimed that the Thingley–Salisbury route would provide communication between Southampton and Bristol 'superior in every respect to any railway which could be made between Basingstoke and Swindon'. Originally, incidentally, the Company had been proposed as the Wilts & Somerset only, Weymouth being added after the Bristol & Exeter – whose Yeovil branch was also authorized in 1845 – had refused to take interest in the port.

Various amendments were made in 1846. For example, the suggested junction between the Salisbury and Weymouth lines was moved from Upton Scudamore to Westbury, to which the Berks & Hants (from the GWR at Reading) was to be extended; and the Bradford branch was to be extended to meet the GWR at Bathampton near Bath.

Page 165 'The withered arm'

Plate (33) Three locomotives on a pre-war Ilfracombe train near Braunton.

Plate (34) A post-war scene at Meldon Viaduct near Okehampton with a 'West Country' class locomotive.

Plate (35) An LSWR *North Cornwall express leaving Padstow, c. 1914.*

Page 166 Three early engravings from the '*Stranger's Guide to Plymouth*'
Plate (36) The construction of the Breakwater (note rails laid on it).
Plate (37) The Plymouth & Dartmoor Railway's tunnel.
Plate (38) Plympton station on the South Devon Railway.

Page 167 Plymouth memories

Plate (39) The 'Illustrated London News' picture of the first train arriving at Millbay.

Plate (40) Changing the gauge outside Millbay in 1892.

Plate (41) The first suburban rail-motor at Saltash.

Page 168 Cornish stations

Plate (42) Bodmin Road.

Plate (43) Bugle, in the china clay district on the Newquay line. Since rebuilt with island platform.

Plate (44) Penzance in the 1950s.

The 53 days' proceedings before the 1847 Parliamentary Committee considering the various direct Exeter railway schemes (see Chapter III) roused almost as much interest in the Wiltshire–Somerset as the Devon–Dorset country. The Wilts, Somerset & Weymouth was affected in two ways. Firstly, a section of its Weymouth line, east from Castle Cary (to which the Bristol & Exeter was to extend) formed part of one of the direct routes authorized in 1848, but of course not built. Secondly, the Company had come to look for considerable traffic from Yeovil, and its supporters were naturally frightened by the 1848 Acts for two narrow-gauge lines to that town. The fact that reaction to the Railway Mania prevented either of these coming to fruition was cold comfort, for the broad-gauge Company, too, was hurt by the depression. Calls were unpaid on a large proportion of shares, and the construction work – started promptly after the passing of the Act – had to be abandoned when ready money was exhausted. Thousands of labourers anticipating a few years' steady employment were turned adrift. Only the first 14 miles, from Thingley Junction to Westbury, were finished. Opening date for the double track, involving but slight engineering works and no serious gradients, was 5 September 1848. The remaining 120 authorized miles were left punctuated by earthworks and bridges in various stages of building. Some of the land was relet.

The branch from Staverton (soon renamed Bradford Junction) north of Trowbridge, to Bradford, however, was so nearly ready that it remains a mystery why it was not finished and opened. Bradford station was waiting and little but laying the rails remained to be done when labour was suddenly withdrawn, to the fury of the town where the large woollen business was beginning to decline seriously. It was nine years before a passenger train entered the station – nine long years, during which much business was diverted to Trowbridge.

The Wiltshire woollen industry generally declined later than Devon's, particularly Exeter's. Helped by the main line, Trowbridge succeeded in keeping its mills until toward the end of the century, when it changed fairly painlessly into a market and administrative centre. As the seat of the County Council, it is still prosperous, if unattractive. Melksham lost heavily, both to Trowbridge and 'up country', when the wool trade ceased, but with the help of a little industry has more than recovered. Not so Westbury, once a thriving market centre, now virtually a village. Its station was too distant to allow much benefit from junction business in the days before the majority of passengers catching trains arrived by car. As in some other parts of the West, one of the railway's first effects was to stimulate population exodus: the old order rapidly crumbled, even in some of the outlying villages.

The Wilts, Somerset & Weymouth had been formed in the belief that a local company would make better financial headway than the GWR itself. But as little resulted, Paddington absorbed the baby on 14 March 1850. Now came a slight reawakening, and single lines were opened from Westbury to Frome, on 7 October 1850, and from Westbury to Warminster on 9 September 1851. A branch for coal traffic from Frome to Radstock was begun. (It was eventually opened on 14 November 1854.)

But the GWR finances were strained and again activity ceased. In 1852 another subsidiary Company, the Frome, Yeovil & Weymouth, was formed in the hope of stirring local interest, but failed. As its Act stipulated that all capital should be subscribed within three months it was dissolved automatically, responsibility for the completion of the whole of the ex-Wilts, Somerset & Weymouth system returning to the GWR proper. Agitation for this completion rose in crescendo. The Mayor of Salisbury complained bitterly that the GWR found it so profitable carrying passengers between Bristol and his city *via* Reading that it lacked interest in the Bathampton–Salisbury route.

Great excitement reigned when proceedings were taken for writs of mandamus to compel the completion of all the lines. Naturally the GWR was made scapegoat for the district's economic troubles, but the Company proved that times were hard for it too, and was successful in its defence – except for the Bradford to Bathampton section. Mandamus for this was made absolute by the Queen's Bench in Michaelmas Term 1853 after a much-publicized trial at Somerset Assizes, and an appeal was rejected.

Next year the Company decided to renew powers for the extensions to Weymouth and Salisbury as well as Bathampton. The price of this move to protect the territory from the narrow gauge was the usual section inserted in the Act – of 31 July 1854 – threatening the suspension of the Company's dividends if the works were not ready within two years. Once more navvies began congregating in the area. By 1856, we are told, the GWR had already spent £1,433,000 on the Wilts, Somerset & Weymouth; and the opening of the rest of the system, through thinly-peopled country, was expected to bring an operating loss. This was but one factor then contributing to the GWR's difficulties and dissension on the Board.

Warminster to Salisbury and Frome to Yeovil were opened as single lines, equipped with the electric telegraph to control crossing movements, on 30 June and 1 September 1856. The Salisbury terminus was beside the site of the present Southern Region station, built several years later; the extension from Yeovil took a few more months, Weymouth eventually being reached on 20 January 1857. Bradford station, built nearly a decade before, was used at last on 2 February 1857, opening day for the single track through the pretty valley from Bradford Junction to Bathampton. Unlike the Westbury–Salisbury and Westbury–Yeovil lines, here engineering works were considerable, including seven viaducts and two aqueducts for the Kennet & Avon Canal.

The Devizes branch, from Holt Junction, between Trowbridge and Melksham, was opened on 1 July 1857 – 'a full week earlier than had been expected so that there was no time to organize the customary celebrations'. But a landslide into one of the cuttings on the steeply graded single line had considerably delayed progress and it was no less than twenty-one years since Devizes had first been included in a railway scheme – the original Great Western from London to Bristol. The town being dissatisfied with branch status, there followed a complicated wrangle for a direct route to London. Schemes included one advanced by the Kennet & Avon Canal (already robbed of much of its traffic) for a railway on its banks. Devizes and its railways could fill a chapter; it will suffice to say that on 11 November 1862 the Berks & Hants Extension was opened, filling in the missing 24½ miles to Hungerford, already reached from Reading.

As early as 1864 the GWR had planned to lay a narrow-gauge rail from Bristol to Salisbury, but this came to nothing. Various other proposals included a curious one for a separate narrow-gauge rail beside the broad gauge to the LSWR at Yeovil; but all that needs noting is the GWR's 1870 agreement with coal-mine proprietors and the Bristol & North Somerset Railway to provide the narrow gauge between Radstock and Salisbury. In view of this obligation the directors announced in February 1874 that they were 'of opinion that the convenience of the public will be best met, and the interest of the Proprietors best secured, by the alteration from Broad to Narrow Gauge of all the Lines in the district', at an estimated expense of £290,000 plus £70,000 for rolling stock. The first figure included an allocation for doubling the rest of the track: Westbury–Frome and Bradford Junction–Bathampton, incidentally, had now been doubled.

GWR platelayers were massed into the district at the beginning of June – still 1874. The last day of the broad gauge between Weymouth and Frome was 18 June, and on 21 June the final train left Frome for Thingley Junction. Single-line working was used on this section until 25 June, when narrow trains began using both tracks. 131½ miles, plus sidings, had been converted in a few days. Doubling followed slowly. The last section between Frome and Yeovil was completed in 1881, and between Westbury and Salisbury in 1901.

'*Via* CASTLE CARY'

Back in 1848, we have seen, a plan was produced for a direct Exeter route using part of the Wilts, Somerset & Weymouth system; so the idea of '*via* castle Cary' was far from new when towards the end of the century the GWR rightly felt it intolerable to continue sending all trains from Paddington to South West England by Swindon and Bristol.

The first step in developing the new route to Taunton was the doubling of the Berks & Hants Extension from Hungerford to what is now known as Patney and Chirton. From a junction there, a new 15-mile double line was built down the Stert Valley through Lavington to Westbury. Opening date was 29 July 1900 for goods and 1 October for passengers. Westbury station, hitherto a dark trap for the fumes from locomotives and from the local iron mine, which once provided substantial traffic, was now rebuilt in its present form of two long islands.

Then Castle Cary had to be linked with the original main line to the West. A completely new line was needed from Castle Cary to Curry Rivel Junction, a little west of Langport, but for the next four miles, as far as Athelney, advantage was taken of the former Bristol & Exeter's branch, now doubled and raised above winter flood level. Beyond Athelney the branch joins the original main line at Durston, and the old single-track route, *via* Lyng Halt, is still used by local trains. But a new double line, by-passing Durston, was built from Athelney to Cogload Junction, about two miles from Taunton.

Opening dates were: Castle Cary to Charlton Mackrell, at first as a single line, 1 July 1905; Curry Rivel Junction to Somerton 12 February 1906 (Langport was given a main-line station but retained that on the Yeovil branch); Charlton Mackrell to Somerton, at first single and for local goods only, 20 May 1906; Athelney to Cogload Junction, 2 April 1906, at first for goods only. Through goods traffic from London to the West began on 11 June, and on 2 July, still 1906, passenger trains started taking the new route, doubled throughout. Altogether 33 miles of new railway were built and 23½ miles of existing single line rebuilt and doubled to bring Taunton 20¼ miles nearer Paddington than before. An army of men was employed, the hourly wage being 6d, reduced to 5¾d despite strong protest.

Engineering works were heavy, including the 1,056-yd Somerton Tunnel and ten-arch Langport Viaduct, with its foundations sunk below 50 ft of peat and bog. Gradients had to be steeper than *via* Bristol, but the absence of congestion more than compensates for this. By-passes at Westbury and Frome, 2¼ and 2 miles long respectively, were opened for goods traffic on 1 January 1933, passenger trains following after the earth works had settled. So today such expresses as the *Cornish Riviera* pass through only one even moderately important station (Newbury) all the way from Reading to Taunton.

Both 1900 and 1906 brought major time-table reshuffles. The opening of the Lavington line (down the Stert Valley) shortened the GWR's Weymouth route by 14½ miles, while Devizes, which had enjoyed semi-main-line status since 1862, was to its chagrin relegated to a branch position again, though for many years a few fairly fast trains including a Paddington–Weston-super-Mare service travelled by it. Those for Bristol took the northern curve (opened on 10 March 1895) of the triangular Bradford Junction.

Closure of the Devizes section took place on 18 April 1966. The Lavington line robbed also the original Wilts, Somerset & Weymouth from Thingley Junction of much business. Though remaining double, Thingley Junction to Bradford Junction then carried little more than local traffic, the only through services using it being those between Wolverhampton and Weymouth, latterly confined to summer Saturdays. Regular services were withdrawn at Melksham as at Devizes on 18 April 1966, but a summer Sundays Swindon–Weymouth excursion continued to use the route, now singled, until 1971. Some freight is still sent this way, and occasionally if there has been a breakdown elsewhere passengers find themselves traversing the original route by which milestones showing the distance from Paddington are still calculated further west.

The Bath–Westbury section *via* Bradford-on-Avon and Trowbridge has always been the busiest of these cross-country lines, with heavy freight traffic and for many years a substantial through service of South Wales and Bristol to Portsmouth and Brighton trains, plus the Bristol–Weymouth service and locals, some of which started or terminated at Trowbridge; this was for a long time a surprisingly busy station considering it was not on a direct route to or from London. This section still remains double and reasonably busy with about a dozen daily passenger services each way.

Nearly all daytime trains from Paddington to the South West were switched to '*via* Castle Cary' in 1906. As well as helping Devon and Cornwall, this valuably reduced pressure on the Bristol line. But the 1906 changes made little difference to the local traffic pattern. Not one of the '*via* Castle Cary' expresses calls at Frome, Bruton or Castle Cary itself, and the intermediate stations from Castle Cary to Langport, served by locals, never justified themselves. They were closed on 10 September 1962.

1906 was too late for a railway drastically to alter the economy of a rural area – not that the erstwhile section of the Yeovil branch had made much difference to the willow-growing district even during the last century. The scenery and way of life in this flat country have changed less than anywhere else in the West, except moorland above the 1,000-ft contour. Except when the occasional express thunders across, carrying passengers for destinations which to the local inhabitants are mere names, there hangs an almost uncanny silence and stillness, perhaps broken by the flapping wings of a heron which has followed the Parrett – the Parrett which until flood-prevention schemes were effected in the 1960s each winter overflowed to turn the surroundings into a vast lake. The expresses have as little to do with the area as the aeroplanes overhead.

When this book first appeared in 1958 the expresses came at two-hourly intervals each way, and in the early 1960s there was talk of abandoning the route, sending all Paddington–Plymouth services *via* Bristol once again; or alternatively singling it. Happily wiser counsels have prevailed, and today there is an hourly train *via* Castle Cary from Paddington to the West from 7.30 a.m. to 7.30 p.m. on Mondays to Fridays with two advertised extras for the weekend rush on Fridays throughout the year and still a dense service on summer Saturdays. Only about half the trains call at Westbury and none anywhere else in the area covered by this chapter, but even this has helped; businessmen from a wide area use Westbury as their railhead, while quite useful connections are made there for Trowbridge and the Weymouth line trains calling at Frome (a thriving town), Bruton and Castle Cary. Freight traffic is also heavier than anticipated some years ago; very heavy cement traffic leaves the large Westbury works, while the stone traffic coming off the stub of the erstwhile Wells and Yatton branch at Witham has already been mentioned in Chapter II. Incidentally, three of Wiltshire's famous White Horses carved into the chalk downs may be seen from the carriage window, including Westbury's, the most celebrated and supposed to commemorate King Alfred's victory over the Danes at Ethendune. It was recut and lined with cement at an expense of £4,000 in 1957. A large-scale map and guide book may be used to advantage throughout this line, many historical and legendary places being visible – and again there being no town larger than Newbury all the way from Reading to Taunton.

Salisbury Plain, once famous for sheep, is today a military centre. The Westbury–Salisbury line's traffic quickly increased as the result of the War Department's purchase of a large tract of land for training purposes in 1895, and between 1914 and 1918 troop movements were phenomenal. Most of the intermediate stations were provided with extra sidings and loading bays, and two branches were built by the War Department. The Codford Camp and Sutton Veny Camp lines were $2\frac{3}{4}$ and $3\frac{1}{2}$ miles long respectively; taken over by the GWR in 1918, they were later closed. Military traffic at Warminster remains considerable, but all stations between there and Salisbury, a stretch of 20 miles, were closed to passengers on 19 September 1955. Since then the through South Wales and Bristol to Portsmouth and Brighton service has declined and most of the present nine daily trains run from Bristol or Westbury only to Salisbury. The GWR's Salisbury terminus was closed from 12 September 1932; already most services had been transferred to the Southern station.

YEOVIL AND OTHER LINES

Though we have seen that part of it was converted into the main line, the Bristol & Exeter's YEOVIL branch may best be included in the rest of Yeovil's complicated story. Durston to Martock was ready in 1849 but lay rusting until the whole broad-gauge branch from Durston to Hendford, on Yeovil's outskirts, was opened on 1 October 1853. On 2 February next year the line was extended through Yeovil to the Wilts, Somerset & Weymouth's Pen Mill station, on the opposite edge. As already said, the Salisbury & Yeovil was opened to Hendford on 1 June 1860, but the Yeovil-Exeter section quickly followed and Yeovil Junction was established on the main line. Yeovil Town joint was opened on 1 June 1861, after which Hendford was used for goods only. The LSWR reached the new station by a curve direct from Yeovil Junction. The original curve towards Salisbury was retained for several years, but has long been abandoned.

Also as already said (Chapter II), the Bristol & Exeter mixed the gauge of its branch as far as Yeovil Town in 1867 but did not run narrow passenger trains until the third rail was extended to Pen Mill in November 1868. Gradually narrow trains outnumbered broad ones on the branch, and on 30 June 1879 the gauge was changed from mixed to narrow. Trans-shipments from gauge to gauge now ceasing at Hendford, the LSWR no longer needed its independent line beyond Town and converted it into a siding. Such were the complications that originally Yeovil Town had four signal boxes, Bristol & Exeter and LSWR at each end, and the platform staff was divided into three – joint, Bristol & Exeter and LSWR.

Yet another Yeovil station, Clifton Maybank, on the down side at Yeovil Junction, was purely a goods transfer point between the GWR and the LSWR. Reached by the GWR by a spur from Pen Mill running under the main Waterloo–Exeter route, it was used

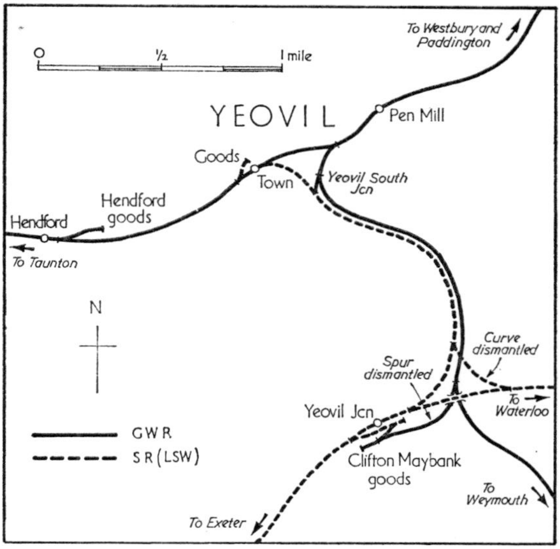

between 13 June 1864 and 7 June 1937, after which goods exchanges worked *via* Town. Passengers from Weymouth going to the Southern similarly travel by Town, changing there and in most cases also at Pen Mill and Junction.

A glance at the map, if not essential to interpret these explanations, will show the wastefulness of the arrangements. One central station would have encouraged more traffic, though all three passenger stations remained busy throughout the 1950s. Both routes had good London services, the Southern supplying an occasional through one from Yeovil Town. Early morning business traffic to London grew steadily after the 1939–45 war. Even the Taunton line carried a fair number of passengers.

The Taunton–Yeovil passenger service was withdrawn on 15 June 1964 and the branch closed completely soon after: that is to say the Durston–Yeovil Town section except for the few miles converted into part of the Castle Cary route from Paddington to Penzance in 1905. A few passenger trains continued to run between Town and Penn Mill until the closure of that section on 29 November 1965; the Town to Junction branch should have closed the same day but because the Traffic Commissioners refused to license a replacement bus service until road improvements had been effected lasted until 2 October 1966. Now no connection is provided between the two routes at Yeovil, and for instance Exeter–Weymouth passengers have to be routed *via* Westbury.

Even now one feels that it might pay to close Penn Mill and Junction and run all trains in and out of the centrally-situated Town, for Yeovil is a boom town. Its trade and population have trebled since the railway arrived, although the railway has not played so important a role as at say Taunton. Luck and local initiative have been important features. It should be added that both routes serving the town are now mainly singled, and while the Exeter–Yeovil Junction–Waterloo service remains of a reasonable standard only diesel multiple units use the Westbury–Yeovil–Weymouth section.

As already suggested, the Bristol to FROME branch was built in two parts. The Frome–Radstock colliery line, opened on 14 November 1854, had a triangular junction at Frome to permit direct working either up or down the Wilts, Somerset & Weymouth. The Bristol–Radstock section was also made to tap coal traffic: the narrow-gauge Bristol & North Somerset Railway was authorized in 1863 but had great difficulty in raising money and did not succeed in opening until 3 September 1873. Until the mineral line to Frome was converted in June 1874 there was a break of gauge at Radstock. From the start passengers were carried on the northern section, but not south of Radstock till 5 July 1875. The Bristol & North Somerset was amalgamated with the GWR in 1884, and the complete branch treated as one unit. It remains single, with a short double section through the colliery area just south of Radstock. The passenger service was more than halved by economy cuts made by the Western Region in June 1958; the trains which remained were then less patronized and passengers ceased to be carried from 2 November 1959.

Freight and mineral traffic continue, but the North Somerset coalfield is fast declining. Figures quoted by the National Coal

Board are 1,494,000 tons mined and 10,000 men employed in 1900; 701,000 tons mined and 3,000 men employed in 1956. More of the coal was then being moved by road than in the Midland and Northern fields; the Bristol–Frome line was carrying about 150,000 tons annually. All but one of the collieries directly served by the line have been closed, but the coalfield's one bright star, Writhlington colliery, on the now-closed Somerset & Dorset line, has had its traffic transferred by a new connection between the two routes west of Radstock stations and level crossings on 7 March 1966 – exactly where such a link had been shown on the S & D Bath Extension Act of 1871! The Bristol–Frome line has been taken up north of Radstock whose industrial archaeological complex of two stations with level crossings is thus served only from Frome, and though Writhlington colliery is performing well now, total closure of the Somerset coalfield is likely within a couple of years, when the branch from Frome will probably be cut back just to serve the stone quarry lines still active in the Frome–Mells Road district.

CAMERTON, like Radstock, was placed on the railway map because of coal. The 3¼ miles from Hallatrow were opened as a branch of the Bristol & North Somerset on 1 March 1882. Owing to steep gradients, goods trains were limited to 15 loaded trucks. Camerton to Dunkerton Colliery, at first used by goods only, followed on 26 August 1907, and Dunkerton Colliery to Limpley Stoke on the Trowbridge–Bathampton line on 9 May 1910, when passenger trains ran through from Camerton. Some of the twentieth-century part of the branch replaced the old Somersetshire Coal Canal, whose 66-yd Combe Hay Tunnel, built in 1805, was used intact. The railway's coal traffic was once considerable, but the passenger service was withdrawn on 22 March 1915. It was restored after pressure from local people on 9 July 1923, only to be ended finally on 21 September 1925. All traffic ceased between Hallatrow and Camerton, and the track was removed. A few years later Dunkerton Colliery, the largest of several on the line, closed, and on 15 April 1950 the remaining colliery, Camerton, followed suit. The last general goods ran on 14 February 1951, after which the only trains were the remarkable assortment of period pieces used for the filming of *The Titfield Thunderbolt* in 1953. The track was removed in 1958.

CHAPTER IX

The Dorset Coast

TO POOLE, BOURNEMOUTH AND DORCHESTER

Every Monday to Friday evening for many years two trains left Southampton Terminus in the same direction within a few minutes of each other and arrived at Wimborne from opposite directions, though still within a few minutes of each other. One had traversed a great built-up area and the other taken a short-cut through the forest. It was not surprising that strangers were confused by the railway geography of the Hampshire-Dorset border country, although with recent closures it is simple enough now.

Though it is outside this book's area, we must begin by mentioning Wimborne, for the small town on the heath's edge was once 'the most important railway centre in Dorset . . . the key to the metropolitan connections of the county'. The station, deserted most of the day – the two from Southampton were among the very few through trains still using the line – was noisy with activity.

This was due largely to A. L. Castleman, the local solicitor who energetically headed the promoters of the Southampton & Dorchester Railway. As was seen in the section on 'An Epic Parliamentary Battle' at the beginning of Chapter III, the line was intended as the first part of a trunk route on to Exeter; but after a long, complicated fight between the Coastal and the Central parties, the narrow gauge eventually reached Exeter *via* Salisbury. The Southampton & Dorchester was thus left dependent on the Dorset traffic, and eventually much of this was diverted through Bournemouth away from Wimborne. Bournemouth, of course, is the clue to such changes: its rise after the initial railway system's completion brought untold complications. The map shows that the same is true of roads, the main network still being based on

THE RAILWAY TIMES.

OPENING of the SOUTHAMPTON and DORCHESTER RAILWAY.—The RAILWAY from SOUTHAMPTON to DORCHESTER, in connection with the London and South-Western, will be OPENED for PUBLIC TRAFFIC on TUESDAY next, the 1st of June, when the trains will run as follows :—

DOWN TRAINS.

Class.	Leave Nine Elms.	Arrive at Southampton.	Leave Southampton.	Arrive at Dorchester.
1st and 2nd ..	..	..	7.50 a.m.	11. 5 a.m.
1st, 2nd, & 3rd	7.30 a.m.	11.30 am	12.15 —	3.30 p.m.
1st and 2nd ..	9. 0 a.m.	12. noon		
1st and 2nd ..	11. 0 a.m.	1.50 pm	2.30 p.m.	5.30 p.m.
Express 1st and 2nd..	12.30 —	2.15 pm		
1st and 2nd, and 3rd c. between Southampton and Dorchester ..	3. 0 p.m.	5.50 p.m.	6. 0 p.m.	9.15 p.m.
Mail 1st & 2nd	8.50 p.m.	12.17 nt.	12.30 nt.	3.30 a.m.
ON SUNDAYS.				
1st and 2nd ..	..	..	7.50 a.m.	11.50 a.m.
1st, 2nd, & 3rd	7.30 a.m.	11.30 am	1.15 p.m.	4.30 p.m.
1st and 2nd ..	10. 0 a.m.	1. 0 pm		
1st, 2nd, & 3rd	..	..	6. 0 p.m.	9.15 p.m.
Mail 1st & 2nd	8.50 p.m.	12.17 nt.	12.30 nt.	3.30 p.m.

UP TRAINS.

Class.	Leave Dorchester	Arrive at Southampton.	Leave Southampton.	Arrive at Nine Elms.
1st and 2nd (and 3rd c. between Dorchester and Southampton ..	7.30 a.m.	10.45 a.m.	11. 0 a.m.	1.50 p.m.
1st and 2nd ..	11.30 a.m.	2.45 p.m.	3. 0 p.m.	5.50 p.m.
1st, 2nd, & 3rd	11.30 a.m.	2.45 p.m.	4. 0 p.m.	8. 0 p.m.
1st and 2nd ..	2.50 p.m.	5.45 p.m.	6. 0 p.m.	8.50 p.m.
1st and 2nd ..	6. 0 p.m.	9.15 p.m.	..	..
Mail 1st & 2nd	10. 5 p.m.	1. 5 a.m.	1.15 a.m.	11.25 a.m.
ON SUNDAYS.				
1st, 2nd, & 3rd	7.30 a.m.	10.45 a.m.	..	..
1st, 2nd, & 3rd	12.30 —	3.45 p.m.	4. 0 p.m.	8. 0 p.m.
1st and 2nd ..	6. 0 p.m.	9.15 p.m.	..	..
Mail 1st & 2nd	10. 5 p.m.	1. 5 a.m.	1.15 a.m.	4.25 a.m.

the turnpikes. Two lesser factors also contributed to Wimborne's decline as a railway centre. As will be seen in the next chapter, the Somerset & Dorset at first made its junction there but then forsook it for Broadstone, while the Salisbury & Dorset Junction line from Salisbury was once far more important than during its last years.

There being no point in repeating the details given in Chapter III, here we are concerned with the Southampton & Dorchester purely as a local line. Indeed, as it wound among the heath-clad hills, avoiding river estuaries and tapping the slight population then in the area, it scarcely appeared possessed of anything but local importance. Castleman's Corkscrew or The Water Snake, certainly; but it did have easy gradients. Lack of heavy engineering works enabled quick construction. The 60½ miles from Southampton *via* Ringwood, Wimborne and Wareham to Dorchester, and a short branch to Poole (or what is now Hamworthy Goods) were authorized on 21 July 1845, and opened throughout on 1 June 1847. Next year the LSWR absorbed the local Company, Castleman being given a seat on the Board. Though at first the five daily trains in each direction were lightly loaded, receipts soon passed the figure (£65,000 annually) at which the Act stipulated the track should be doubled. The first section was finished in 1857, and the entire run in 1863.

The Act gave running powers from Dorchester to Weymouth over the broad-gauge Wilts, Somerset & Weymouth, but as the latter was making scant progress a separate route to the Royal resort was planned. However, the original arrangement was cemented by another Act, in 1854, and LSWR trains were extended to Weymouth on the same day as GWR ones, 20 January 1857. In return for providing narrow rails on this section, the GWR insisted on broad ones on the LSWR for the first eight miles east of Dorchester towards Wareham. Whether Paddington was merely out for a pound of flesh, however useless, or whether it hoped to use the eight miles as the start of a broad route on to Salisbury or Newbury remains a mystery; at all events the rails petered out in the blasted heath at the end of the prescribed distance and were never used!

Thomas Hardy's Egdon Heath has changed considerably since Tess of the D'Urbervilles spent her tragic honeymoon night in the manor house whose original still stands near Wool station. More land has been cultivated and some given to conifer plantations; but pieces of open, windy heath still sweep down to the track. Incidentally, the railway's role in the life of a sparsely populated, semi-barren country in the middle of last century is nowhere better described than in Hardy's novels; though perhaps they do not fully bring out the devastating effect on main-road villages by-passed by the trains. Road traffic declined even more sharply in Dorset than elsewhere upon the advent of the steam age. Old-established markets faded as new ones were opened nearer the railway; public houses once thronged with travellers, including seamen crossing by land from Weymouth to Bristol to save going round the Cornish coast, were left dependent on the local population, itself declining and impoverished. Smiths and other craftsmen, shopkeepers and carriers, fared no better. Village carters and merchants who had previously fetched coal and other goods from one of the coast's small harbours now made the much shorter journey to the nearest station. Their one consolation was that despite the declining population demand for coal remained steady, owing to its lower price by rail and the gradual introduction of machinery, including steam ploughs. All this, of course, was not the result of just the Southampton–Dorchester line, but of the county's whole railway system over a long period.

A story often repeated is that of the growth of Bournemouth: it is, indeed, almost miraculous. When Castleman's Corkscrew was opened, the place was a coastguard hamlet of about 30 houses, dating back only to 1810. Even neighbouring Poole, with a population of about 6,500 was not thought to warrant a branch to its centre – though it was a busy port and had a rising pottery trade to compensate for the loss of previous sources of revenue. (About 60,000 tons of clay were being raised annually in the surrounding district.) Passengers to the town travelled down what is now the Hamworthy Goods branch and crossed by a toll bridge. When Bournemouth began attracting visitors they had to use the same roundabout route. One of them wrote:

> I began with a 15 mile drive to Chippenham, and then my troubles began and after exploring what seemed to me the whole of the South Coast, I finally reached Hamworthy. A porter informed me Bournemouth was five miles further. He

had never been there but there was a good road and I had better go by Oxford . . . The mystery was solved by the driving up of a little old omnibus, owned by 'G. Axford', who conveyed passengers once a day to Bournemouth.

In 1870 the resort, whose population was then nearing 6,000 (ten years later it was 16,000), was reached by a branch from Ringwood *via* Christchurch (see Vol. II). Land productive of little but pine-trees shot up in price as boldly planned streets and crescents took shape. The pines, as well as beautifying the valleys or 'chines', were supposed to be 'salutiferous'. Poole gained both by its neighbour's growth and on its own account. The Hamworthy branch had become insufficient as early as 1863, when through trains from Burnham-on-Sea began running *via* the Somerset & Dorset. So in 1865 a Poole & Bournemouth Railway obtained powers for a line from Poole New Junction (today's Broadstone) to Poole proper and then 'to the favourite watering place of Bournemouth'. It was largely a Somerset & Dorset project but owing to that Company's poverty was aided by the LSWR, which worked and eventually absorbed it. The line, at first single but doubled in 1883, was opened from Broadstone to Poole on 2 December 1872, and to Bournemouth West, including a sharply graded section up from Parkstone to Branksome, on 15 June 1874. Traffic naturally declined on the original so-called Poole but in fact Hamworthy branch, which had been doubled with the rest of the Southampton & Dorchester; it was now singled, and on 1 July 1896 was closed to passengers.

At first the two railways to Bournemouth (*via* Ringwood to East station and *via* Poole to West) were unconnected. But Bournemouth Central replaced East on 20 July 1885, and from 6 March 1888 trains continued from there over the LSWR's Bournemouth Junction line to West. (The same day the LSWR opened a shorter route from Southampton, *via* Sway.) If the resort had grown quickly while access was still difficult – some of the élite fought tooth and nail to preserve select isolation – now that it had a good train service the rate was phenomenal. Some extra carriage accommodation became essential, and the most economical way to supply that was to divert the Weymouth portion of expresses through it. At this time the Bournemouth and Weymouth sections were parted at Brockenhurst. Often the Weymouth

RINGWOOD, CHRISTCHURCH & BOURNEMOUTH EAST LINE.

All Trains are First, Second, and Third Class.

	WEEK DAYS.												SUNDAYS		
					Fast			Fst		Fast					
Waterloo ... dep.	...	...	5 50	8 5	11 15	11 15	...	2 15	3 15	4 55	5 50	Thursdays only.	..	10 15	...
Clapham Junc. „	...	...	6 1	7 57			...	...	...	...	...		...	10 26	...
Surbiton „	...	...	6 16	8 26	11 36	11 36	...	...	...	...	...		...	10 42	...
Woking Junc. „	...	...	6 44	8 48			...	...	3 54	.	6 30		...	11 8	...
Basingstoke ... „	...	...		9 39	12 35	12 35	...	3 28	4 37	6 5	7 13		..	12 5	...
Winchester ... „	.	...	8 24	10 15	1 6	1 6		3 55	5 8	6 33	7 49		..	12 40	...
Bishopstoke ... „	...	..	9 0	10 35	1 23	1 32	..	4 7	5 30	6 46	8 8		..	12 57	...
Portsmouth ... „	...	5 20	8 0	9 30	...	12 35	...	2 15	4 30	...	8 43		...	8 50	...
Southptn. Wt. „	...	6 41	9 15	10 50	...	1 44	..	4 19	5 43	6 58	8 21		7 41	1 37	...
Redbridge ... „	...	6 50	9 24	10 57	..	1 50	...	...	...		8 29		7 51	1 14	...
Lyndhurst Rd. „	...	7 0	9 35	11 8	...	2 1		...	5 59		8 40		8 6	1 56	...
Ringwood ... arr.	...	7 37	10 11	11 46	2 10	2 38	..	4 58	6 33	7 36	9 14		9 0	2 33	.
Ringwood ... dep	...	7 48	10 20	11 56	2 12	2 45	...	5 8	6 40	7 40	9 17	...	9 5	2 40	4 40
Herne ... „	...	8 7	10 34	12 10	...	3 0	...	...	6 54	7 53	9 31	.	9 19	2 54	4 54
Christchurch „	5 45	8 15	10 42	12 18	2 33	3 8	5 0	5 24	7 2	8 1	9 39	10 15	9 27	3 2	5 2
Brnmth., E. arr.	5 53	8 23	10 50	12 26	2 41	3 17	5 8	5 32	7 10	8 9	9 47	10 23	9 35	3 10	5 10

	WEEK DAYS.											SUNDAYS		
			Fast		Exp.									
Brnmth, E. dep.	5 30	7 10	8 55	9 35	11 0	1 20	3 40	4 40	5 40	7 30	10 0	7 20	2 0	3 55
Christchurch ... „	5 38	7 18	9 3	9 43	11 9	1 28	3 49	4 48	5 48	7 38	10 8	7 29	2 8	4 3
Herne ... „	...	7 26		9 51		1 36	3 47	...	5 56	7 51		7 38	2 17	4 12
Ringwood ... arr.	...	7 59	9 22	10 5	11 20	1 50	4 11		6 10	8 4	...	7 49	2 30	4 25
Ringwood ... dep.	...	7 44	9 28	10 15	11 33	1 58	4 19	.	6 21	8 12	Thursdays only.	7 54	...	4 29
Lyndhurst Road „	...	8 18	...	10 55	...	2 25	4 54	...	6 58	8 48		8 28	...	5 7
Redbridge ... „	...	...	...	11 6	.	2 46		...	7 7	8 59		8 38	...	5 20
Southpton, Wt. „	...	8 34	10 6	11 17	12 18	3 53	5 14	..	7 15	9 10		8 49	...	5 32
Southpton Dcks. „	...	...	10 0	...	...			...		9 25		9 30	...	6 0
Portsmouth ... arr.	...	9 52	11 25	12 45	2 19	4 18	6 28		8 59	10 35		10 46	...	9 8
Bishopstoke ... dep.	...	9 1	10 20	11 33	12 56	3 25	5 40	...	7 42	9 41		9 51	...	6 21
Winchester ... „	...	9 16	10 31	11 45	12 40	3 41	5 55	.	7 57	9 56		10 11	...	6 40
Basingstoke ... „	...	9 52	11 1	12 19	1 12	4 15	6 33	...	8 37	10 30		10 48	...	7 18
Woking Junc. ... „	...	...	...	...	...	...	7 5	...				11 40	...	8 3
Surbiton ... „	...	...	...	1 18	..	Sig.	7 23	...	.	11 27		12 3	...	8 25
Clapham Junc. „	...	...		...	...	.	7 37	...	.	11 40		12 22	...	8 38
Waterloo ... arr.	...	11 2	12 10	1 36	2 18	5 35	7 48		9 49	11 51		12 37	...	8 50

WIMBORNE, POOLE, PARKSTONE & BOURNEMOUTH LINE.

THE CONNECTIONS OF TRAINS TO AND FROM WIMBORNE, &c., ARE NOT GUARANTEED.

All Trains are 1st, 2nd and 3rd Class.	WEEK DAYS.																			SUNDAYS					
From BOURNEMOUTH (WEST).						A			A							A			C						
Bournemouth (West) ... dep.	5 20	6 45	7 25	8 30	9 15	10 5	10 30	...	11 45	1 0	...	2 10	3 15	4 0	...	5 15	6 10	8 30	10 30	7 0	3 30	...	...	...	...
Parkstone ... „	5 27	6 52	7 32	8 37	9 22	10 12	10 37	...	11 52	1 7	...	2 17	3 22	4 7	...	5 22	6 17	8 37	10 37	7 7	3 37	...	...	...	...
Poole ... „	5 32	6 57	7 37	8 44	9 27	10 17	10 42		11 57	1 12	...	2 22	3 27	4 30	...	5 27	6 22	8 42	10 42	7 12	3 42	...	...	...	...
Poole Junction and Broadstone ... „	...	7 6	7 46	8 53	9 36	10 26	10 51		12 6	1 21	...	2 31	3 36	4 42	...	5 36	6 31	8 51	10 51	7 21	3 51	...	...	...	...
Wimborne ... arr.	...	7 12	7 52	8 59	9 42	10 32	10 57		12 12	1 27	...	2 37	3 42	4 56	...	5 43	6 37	8 57	10 57	7 27	3 56	...	...	...	...
Wimborne dep. for Somerset and Dorset Line ...	...	...	8 0	...	9 50	10 35	..	12 13	12 25	...	...	2 38	...	5 45	...	...	6 52	9 15	...	..	...	...	...	...	...
Wimborne dep. for London, Southampton, &c. ...	...	7 26	...	9 8	9 55	...	11 11	...	...	1 40	...	...	3 50	...	...	6 5	7 53	11 8	...	7 34	4 6	...	...	...	...
Wimborne dep. for Salisbury, via Fordingbridge ...	...	...	8 35	...	...	...	11 20	...	...	...	...	2 37	...	...	...	...	6 48	...	..	7 45	...	...	...	...	...
Wimborne or Poole Junc. dep. for Dorchester ...	...	...	7 59	...	10 33	10 39	11 0	...	12 10	...	...	3 0	...	5 16	...	5 48	6 52	9 30	1 45	9 24	...	...	...	...	...
To BOURNEMOUTH (WEST).		B						B								C					B				
Wimborne arr. from Somerset and Dorset Line ...	...	...	...	9 20	10 35	...	...	1 27	...	2 58	4 33	5 12	6 35	8 23	9 13	...	...	...	...	...	...	...	...	...	...
Wimborne arrive from Southampton, &c. ...	...	...	7 57	...	10 30	...	12 8	...	...	2 56	...	5 14	6 50	7 52	9 28	...	...	...	...	9 34	8 10	...	...	...	...
Wimborne arr. from Salisbury, via Fordingbridge ...	...	...	7 45	...	...	10 58	...	...	...	2 37	...	5 31	...	8 18	.	...	...	...	...	10 7	..	...	...	...	...
Wimborne or Poole Junc. arr. from Dorchester ...	...	7 20	8 30	9 8	9 48	11 8	...	1 31	1 36	2 30	3 53	...	6 3	7 50	...	11 7	...	...	...	7 33	3 53	...	...	...	...
Wimborne ... dep.	...	7 20	8 30	9 32	10 45	11 15	12 15	1 30	2 5	3 10	4 35	5 33	7 0	8 24	9 37	11 10	...		...	10 10	4 5	...	...	...	...
Poole Junction and Broadstone ... „		7 27	8 36	9 39	10 52	11 21	12 22	1 36	2 12	3 17	4 42	5 40	7 7	8 31	9 44	11 17	...		...	10 17	4 12	...	...	...	...
Poole ... „	5 40	7 37	8 45	9 47	11 0	11 30	12 30	1 50	2 21	3 27	4 50	5 48	7 15	8 41	9 52	11 25	...	...	...	10 25	4 20	...	...	...	...
Parkstone ... „	5 46	7 43	8 51	9 53	11 6	11 36	12 36	1 57	2 28	3 31	4 56	5 54	7 21	8 47	9 5[illegible]	11 31	...	...	...	10 31	4 26	...	...	...	...
Bournemouth (West) ... „	5 55	7 51	8 59	10 1	11 14	11 43	12 44	2 4	2 35	3 39	5 4	6 2	7 29	8 55	10 6	11 39	...	...	...	10 [illegible]	4 34		...	...	...

A Passengers for Down Main Line Train to change at Poole Junction.
B Passengers for Bournemouth West Line, by these Trains, must change at Poole Junction.
C The 10.30 p.m. Train from Bournemouth West to Wimborne and the 11.10 p.m. Train from Wimborne to Bournemouth West will run on Wednesdays only.

To and from Bournemouth in June 1884.

ones carried few passengers, the service being generous for the traffic. The original five daily trains as far as Dorchester had been increased to thirteen even by 1884.

Two new curves were added to enable direct running from Bournemouth Central to Wareham and beyond. The Branksome curve, making a triangular junction north of Bournemouth West, and the Holes Bay curve across tidal water from Poole to Hamworthy Junction, avoiding the necessity of reversal at Broadstone, were both opened on 1 June 1893. The district's railway system was then complete, the only other physical change being the singling of the original Southampton & Dorchester main line between Broadstone and Hamworthy Junction, a section little used for many years before its closure on 4 May 1964.

On the same day the Broadstone–Brockenhurst section of Castleman's Corkscrew was also closed to passengers. Although it remained double track, it had purely local importance except on summer Saturdays when one or two Waterloo–Weymouth expresses used it to bypass Bournemouth and Poole. The latter station, with its two cramped, curved platforms, and two level crossings, has always been difficult to work. Up to 120 trains called there on summer Saturdays in the 1950s, when there was a multiplicity of through services from the north over the Somerset & Dorset and, in addition to the local trains to Brockenhurst and Salisbury, there were even through Bournemouth–Cardiff workings *via* the Salisbury & Dorset Junction line. Now, of course, only the Weymouth line trains remain, although since the closure of Bournemouth West in 1965 the daily through train from Newcastle now makes Poole its terminus (as did the *Pines Express* from Manchester rerouted *via* Reading until it was withdrawn). The closure of Bournemouth West was part of the Waterloo–Bournemouth electrification scheme which, opened on 10 July 1967, also resulted in a better service through to Poole and Weymouth (diesel hauled, although electrification is forecast). Poole's cramped station thus remains a disadvantage, especially to road traffic. Meanwhile the railway era has ended for passengers at Wimborne, and at Broadstone with its large four-platform station. But for several years after the withdrawal of passenger trains, Broadstone curiously remained a railway crossroads, with routes from Poole, Wimborne, Blandford and Upton, the latter being near Dolton's clay pit halfway down the Broadstone–Hamworthy Junction section. The Blandford section closed in 1965, and the clay pit two years later. Freight is still handled at Broadstone and at Wimborne still includes coal traffic. Wimborne is also the place where Trainex Buildings Ltd, the firm with the British Railways' contract for exhibition trains, do their work – involving a total of about twenty-four coaches. Castleman's Corkscrew in fact lives on even beyond Wimborne, two or three trains a week still serving a War Department depot at West Moors. The original Hamworthy branch which gave Bournemouth its very first trains is more active, carrying currently special trainloads of coil steel and cars for export. Its role could increase as Poole Harbour is developed to serve the growing population. Container traffic could be developed.

Returning to Poole, when the line was being built, the Council dropped opposition to the level crossings on a perpetual guarantee that all trains would stop at the stations; many years later when a Weymouth train omitted to call, the agreement was unearthed and its observance insisted upon. Poole's goods traffic was by far the heaviest in Dorset, the town remaining a thriving port as well as industrial handmaiden to Bournemouth, but under the centralization scheme ordinary goods traffic ceased to be handled in 1972, though the mini-marshalling yard (serving Wareham, Furzebrook, Hamworthy, Broadstone and Wimborne) remains, in addition to the cement factory traffic. Marshalling is later likely to be concentrated at Eastleigh and to release more valuable development land the cement works may move to Hamworthy Junction. The ¾-mile Poole Quay line, which ran through the streets, was closed on 2 May 1960.

With a combined population of 300,000, Christchurch, Bournemouth and Poole are virtually one place – almost entirely a railway creation. Continued expansion still depends largely on trains, but increasingly on the electrified multiple unit (and alas not always very comfortable) service to Waterloo, and less on cross-country services than when the first edition of this book was published. In 1958 it was still possible to say: 'Even in winter, for instance, the *Pines Express* from Manchester, *via* the Somerset & Dorset line, brings a reasonable number of visitors – at least as many as go to Torquay by the *Torbay Express* from

London itself.' Today in any case more holidaymakers go by road and the railway's role is largely that of carrying people on business, especially to London. Traffic rose sharply after electrification and shortly there will be a fast, semi-fast and slow train from Waterloo every hour.

TO WEYMOUTH

In a way, Weymouth's rise is even more fascinating than Bournemouth's. It is certainly closer wrapped in England's history; but this is not a history of England or even of holiday resorts and must accept the Royal, fashionable watering-place as a *fait accompli* before the railway age. At first demand for a line was lukewarm. The town had flourished beyond all expectations since Ralph Allen of Bath, improver of the postal system and immortalized as Squire Allworthy of *Tom Jones*, came to restore his health and popularize the sea bathing-machine – and even more since George III's first memorable dip to the accompaniment of the band's rendering of the National Anthem; and like most firmly established places it was slow to appreciate that times were changing. But the benefits of Royal patronage steadily wore away and visitors turned to the new resorts on the railway map. Then the leading citizens clamoured madly for trains, though at no stage were they generous in investing their own money.

It was partly bad luck that the trains did not steam in at least not much later than they did in Plymouth, for many of the Wilts, Somerset & Weymouth's earthworks were cut in 1847 and 1848. The reason for delay in completion was explained in the last chapter, and it has been noted that broad-gauge trains began running from Yeovil to Weymouth on the same day that narrow ones were extended from Dorchester to Weymouth, 20 January 1857. By then the lack of mechanical transport had done the town serious harm; but as in the case of Devizes the precise opening date was finally slightly earlier than expected, and gaiety – somewhat forced – broke forth a week after the six broad and five narrow daily trains had been inaugurated.

The Wilts, Somerset & Weymouth's continuation from where we left it at Yeovil was first single as far as Dorchester, but space was left for a second track and this was laid in 1885. Engineering works were heavy, including four tunnels and seven viaducts, and gradients were unavoidably steeper than elsewhere on the Wilts, Somerset & Weymouth system. The first ten miles from Yeovil are continuously uphill, including two miles at 1 in 51 to Evershot, then a steady fall to near Dorchester, a rise to the last tunnel, and a sharp fall with a section at 1 in 50 to the terminus. This, soon regarded as the ugliest building in Weymouth, was then 'of an exceedingly handsome description, with a light and eloquent roof, under which extend three lines of mixed gauge, and a division to the right provides a place of disembarkation, so to speak, for arrivals by the South Western line'. The station occupied land reclaimed from Radipole Lake or the Backwater: a long-cherished scheme had been to reclaim much more and keep it all as an open space, and the sale to the railway roused strong feelings. The last tunnel, incidentally, Bincombe, 814 yd long, is under the famous Ridgeway Hill, from which travellers on the Roman road caught their first glimpse of the bay. The cutting outside the south entrance revealed a remarkable rock fault, which excited geologists but has involved the railway in worry and expense ever since.

For over a century arrangements at Dorchester were complicated. The Southampton & Dorchester poised itself ready for an extension to Exeter: the original terminus is at right angles to the Western line. When the sharp curve linking the two lines was opened, a new down through platform was provided for Southampton to Weymouth trains, but no corresponding up platform was built and Southern trains therefore backed into the old terminus. Only in 1960 was Dorchester South given a new through up platform. Around the same time a new signal box took over control of electric signalling throughout the Dorchester–Weymouth area.

Dorchester, which would have gained enormously by being on a trunk route to the far West, was of course bitterly disappointed that this was not made into a through station as was originally intended. Even so – even though no new industry came to replace the once all-important cloth manufacture and the coming of the railway resulted in the end of the brewery trade – the place has no reason to complain. In 1851 the population was 3,513 or less, for instance, than Shaftesbury's. In 1971 it was 13,736, between three and four times Shaftesbury's. As a junction town Hardy's Casterbridge quickly stole business from lesser places and has developed almost entirely as a market, administrative and tourist centre. A miniature Exeter, it is highly dependent on the surrounding countryside and its agriculture. At the beginning of the railway era, sheep and crops ranging from turnips to flax were most important, but in common with the rest of Britain crops declined after 1870, and then the sheep population fell drastically. 1879 was a disaster year. But already cattle were coming into their own, and an extra 15,000 were added in the county during the next five years. This transition to dairying, more rapid than in Devon, saved Dorset from the worst effects of depression, though even in nominally good times the farm labourers' lot was harsh. Many, including tenants of small farms, were forced to leave the land and take a single ticket to London or the Midlands. By 1906, we are told, the LSWR alone was sending away 5,000,000 gallons of milk a year from various stations: almost every station, in fact, contributed its daily churns. Today, to the inconvenience of motorists and the underemployment of the railways, much of the milk leaving the county goes by road, but other agricultural traffic is fairly heavy.

At first the LSWR, which maintained its own staff at the station till 1912, had more or less a monopoly of London traffic at Weymouth. While its trains went by Wimborne on a route three miles shorter than the Bournemouth line now used, the GWR's had to go to Chippenham before joining the main line. The LSWR's initial best timing was 4 h. 20 m., by an express carrying 'first class passengers only, with the exception of servants in actual attendance upon their employers'. Incidentally, to begin with the GWR charged 60 per cent of gross receipts in respect of the LSWR's running powers, 'calculated on the rates now chargeable on the LSWR between London and Southampton, exclusive of the cost to them of conveying such traffic between Dorchester and Weymouth'. This odd arrangement was because higher fares were charged on the Southampton line than on the rest of the LSWR.

Local traffic seems to have grown quickly at Weymouth, while right from 1857 some long-distance passengers arrived to embark for the Channel Islands. The port's connection with the Islands dates from well before railway days, but business now became substantial – in the hands of the Weymouth & Channel Islands

Steam Packet Company, closely associated with the GWR. The GWR did not run its own boats into Weymouth until 1876, when a daily service was started jointly with the Western Railway of France. Little more successful than had been a Poole–Cherbourg route in connection with the Somerset & Dorset (see next chapter), it was withdrawn in 1885: in 1883 a ship was totally wrecked. In August 1889 the GWR took over the Weymouth & Channel Islands Company. From 1 July in the same year boat trains ran through the streets down to the quay on the Weymouth Harbour Tramway, part of the Weymouth & Portland Railway, leased jointly to the GWR and LSWR. This had been opened for goods (horse-drawn until about 1885) on 16 October 1865; the gauge was mixed, but the narrow rails were not used until the abolition of the broad gauge in the district in 1874.

Now came a colourful period. Looking forward to developing business when the Lavington line would shorten the distance from Paddington, the GWR began a daylight service to the Islands in 1897. Next year an Act was passed for enlarging the harbour and for a new four-and-a-half-mile line to it, leaving the main line near Upwey Junction. A 1,121-yd tunnel was to be included, but despite much optimism the scheme was dropped and the boat trains still travel through the streets heralded by a porter's warning bell and a bell on the engine. In 1899 the LSWR reminded the public that it still had by far the shortest route from London by substantially cutting its schedules. A spectacular 60 m.p.h. run was introduced between Dorchester and Wareham. As already stated, the GWR improved its timetable on the opening of the Lavington line on 1 October 1900—though still the LSWR had the advantage of some 12 miles.

Although the Southern route remained quicker, the Western retained the lion's share of boat traffic until 1959. Since November that year Channel Island boat trains, whether from Weymouth or Southampton, have run to Waterloo. The shorter crossing from Weymouth, in daylight against the Southampton night departures, won a large proportion of holidaymakers, and today all passenger services are from Weymouth. In summer the boat trains still pass through the streets to Weymouth Quay, but in winter passengers are transferred by bus. The number of passengers embarking at Weymouth has increased threefold this century, despite stiff air competition. Weymouth used placidly to handle such things as 500,000 cwt of tomatoes in a season, but from May 1972 all freight was withdrawn from the Channel Islands–Weymouth route and the town's goods station then closed, the remaining business (mostly coal) being transferred to Dorchester South. The weekly service introduced by the GWR to Nantes in 1909 should be mentioned to complete the picture. It lasted only two years.

Weymouth is smaller than most people expect, but despite the intimacy of its centre, its population has increased from 12,038 (Weymouth, Melcombe Regis and Wyke Regis) in 1851, and 19,416 in 1901, to 42,349 in 1971. The Admiralty connection, of which more when we reach Portland, has helped as well as the port and holiday business. Whether the town would have grown even larger had the railway reached it earlier and provided better cross-country services, is difficult to say, but considering its favourable natural situation the answer is probably yes. The contrast between the cross-country services of Bournemouth and Weymouth is great; yet so too has been that between the methods used by the two resorts to develop their trade. Architecturally, Weymouth owes much to its early origin, but from the business point of view living in the past acted as a continual brake until into this century. Bournemouth started with a clean sheet, free of prejudices and full of enterprise. Not, of course, that a town's mere size or even volume of trade is the only consideration. On the busiest day of 1958, to quote one example, 13,335 passengers (from local and long-distance starting-points) arrived at Weymouth. Bournemouth's total was 16,670, of whom 8,840 arrived at Bournemouth West.

In winter about 30 trains left Weymouth daily in 1958, roughly half Southern, half Western. For some years the Western route as far as Castle Cary was included in the Southern Region – which began installing upper-quadrant signals – but since the 1963 boundary revisions the Western Region has extended as far as Dorchester. In 1973 the total number of weekday departures from Weymouth was down to 25, 16 of them for Bournemouth. Not only boat trains, but all through London services now take the Waterloo route; the best Western trains at Weymouth are semi-fast diesel multiple units to and from Bristol. The track is singled from the junction with the Bournemouth route north through Yeovil though after the first quarter of a mile there is a passing loop at Dorchester West where both platforms remain in use. The importance of the Waterloo–Bournemouth–Weymouth route has increased steadily since electrification to Bournemouth and whether or not the plan to electrify through to Weymouth is carried out (at present four or eight electric coaches through from Waterloo are pulled by diesel to Weymouth and propelled back to Bournemouth), London should effectively be brought closer with a more frequent service of expresses, a big improvement being expected in 1974. The emphasis in the whole area is of course very much on passengers but Wareham retains the curiosity of being starting point for several oil trains weekly; a special kind of oil is drilled from under the sea in Lulworth Cove.

BRANCH AND MINOR LINES

That the Isle of Purbeck, though only a peninsula, retains much island individuality is largely due to the late completion of its only standard-gauge railway, the SWANAGE branch, from Worgret Junction, on the main Southampton–Dorchester route a mile west of the busy little town of Wareham. A single line with fairly heavy engineering works, it was opened on 20 May 1885 and run by the LSWR, which acquired it from the local Company the following year. Swanage, with a population of only 2,000, was then largely dependent on its marble and other quarries, but an increasing number of visitors arrived by the branch, and today it is a thriving little resort with a population of over 8,000. Passenger traffic was relatively heavy when this book was first published; there were several daily through coaches from Waterloo. The only intermediate station was named after Corfe Castle, of which there was a striking view from the train. Since the beginning of the century, clay has formed the heaviest freight, but much of it now goes by road to Poole Harbour. The railway's share, mainly for long-distance destinations, is 30,000 tons annually. Passenger trains ceased on 3 January 1972, a closure many believe should not have taken place, but about 20,000 tons of clay are still collected from Furzebrook, short of Corfe Castle.

Formerly all the clay was taken to three quays on Poole harbour by tramways owned by the clay companies. Primitive rail transport was probably introduced by the industry before Dorset's first main lines appeared, but animal power sufficed until the arrival of the diminutive 0-6-0 tank *Primus* on the 2 ft 8½ in. gauge line from WEST CREECH to a wharf near Ridge in 1866. *Primus* was quickly followed by *Secundus*, used for ninety years and now in a Birmingham museum, and other Emmett-like creations up to *Septimus* followed in due course, a maximum of five engines and 150 ten-ton trucks being used on the 6-mile route at the same time. Branches and sidings brought the mileage up to nearly 20. It was a busy system and in later days holiday motorists on the Wareham–Swanage road were delighted to give way to a train – or in the evening to a jury car freewheeling with a load of workmen for Ridge – on the level crossing. But at the quay the clay could only be loaded onto barges and it had to be transferred again, to sea-going ships, at Poole. It was an expensive method and the War Department, wanting the heath for manœuvres, was only speeding the inevitable by commandeering and closing the line to the east of the Swanage branch early in the 1939–45 war. Much of the clay had already been transferred to the latter line at the point of intersection, and the rest of the narrow system was active until 1956. After 1957 only sidings in the Creech Heath district were left: for a few more years they lingered on, normally worked by a diesel, but *Tertius*, a Manning Wardle 0-6-0 tank of 1886 vintage, remaining as spare and a reminder of the past. Clay from NORDEN went by one of two 3 ft 9 in. gauge tramways, across the barren heath to Middlebere Wharf and to Goathorn Pier. The Middlebere line was closed many years ago, and through traffic ceased to Goathorn in 1936. A mile and a half of track between the pier and a small clay works had been built before the rest of this branch and was left temporarily afterwards. Otherwise only four miles of the 3 ft 9 in. gauge were retained – in the Norden area – and in 1948 this was rebuilt to the more conventional 1 ft 11½ ins. *Russell*, a 2-6-2 tank engine which once hauled holidaymakers over the Welsh Highland Railway, was for a time employed on the line but left Dorset many years ago. Even the shunting diesels have gone, no narrow-guage railways remaining alive in the area.

The BOVINGTON CAMP branch, about two miles long from the main line near Wool, was started during the First World War but not opened until 9 August 1919. Though worked by the LSWR and then the Southern Railway, it was used exclusively for military traffic and closed completely on 4 November 1928.

The Isle of Portland, connected to the mainland only by the Chesil beach, is a solid block of limestone, for centuries quarried intensively. Apart from minor tramways, the first line was the Portland Railway, better known as the MERCHANTS Railway, authorized with 100 shares of £50 in 1825 and opened in October 1826. Of standard gauge, the line ran from 400 ft above sea level on the western side of the Isle, down an inclined plane past the site the prison now occupies, to Castletown Pier. During part of last century, visitors were carried to the prison on Sundays; normally stone was the only traffic, but so considerable that for well over a century the railway prospered and between the wars the shares fetched £150 on the open market. But the last trucks were lowered down the incline in 1939. In 1958 the Company was still not wound up, however.

As well as providing the Isle with trains, the WEYMOUTH & PORTLAND Railway's embankment incidentally strengthened the road link across the shingle to the mainland. Opened on 16 October 1865, the line of heavy engineering works, including a wooden viaduct (replaced by a girder structure in 1909) across the Backwater, had mixed gauge and was leased jointly to the GWR and LSWR. The broad rails may never have been used. At first passenger trains reversed at Portland Junction to reach Weymouth terminus, but Melcombe Regis station, opposite the terminus short of the junction, was opened in 1909 ready for the introduction of an intensive railmotor service (calling at two suburban halts in Weymouth) the following year. Meanwhile the EASTON & CHURCH HOPE Railway, authorized in 1867 to build a line from quarries to Church Hope Cove – it was started but abandoned in 1874 – had obtained further powers and opened from Easton to near Portland for goods on 1 October 1900 and for passengers on 1 September 1902. The gap between it and the Weymouth & Portland was filled by using a section of the BREAKWATER Railway, opened for stone traffic in 1878.

The breakwater, started in the 1840s, was completed in two stages and by the end of the century had turned Weymouth Bay into one of the world's safest anchorages, though not impregnable to sea attack as the firing power of ships increased. After 1918 the Navy reduced activity in the port, one reason for a population decline on the Isle in recent decades, though the total is still above pre-railway figures. The GWR and LSWR (later Southern) jointly worked the Breakwater line (Admiralty property) and the Easton & Church Hope as well as leasing the Weymouth & Portland. Passenger and stone traffic as far as Portland was once very heavy, and fairly so along the cliffs to Easton – there were serious landslides causing suspension of traffic in 1907 and 1935. But the population reduction coupled with better bus services resulted in passenger trains being withdrawn throughout on 3 March 1952. During 1940–4 Easton had been served during the summer only. Freight ceased at both Portland and Easton on 5 April 1965.

ABBOTSBURY has shrunk by more than half in the past century, losing ground particularly after its railway was opened on 9 November 1885 and much mill and other business was transferred to Dorchester and Weymouth. The railway, built by a local Company, absorbed by the GWR in 1896, was never profitable, and when motors became common few visitors used it even on their way to the swannery. Traffic ceased on 1 December 1952, and the track has been dismantled except for a short section used by goods, between Upwey Junction (now Upwey) and Upwey (Upwey Goods) retained until 1962.

BRIDPORT, though several times promised a place on a trunk line to the West, did not see its first train until 12 November 1857. The Bridport Railway, from Maiden Newton, took an indirect course through the hills and was broad gauge until 1871. On 31 March 1884 it was extended to Bridport Harbour – renamed West Bay in deference to the Company, anxious to develop tourism. Most of the harbour's commercial trade had gone – there was a 90 per cent drop between 1830 and 1880 – largely as the result of the failure of Bridport's industries allied to the local flax and hemp growing. But West Bay did not succeed as a resort: passenger services on the extension were withdrawn temporarily during the First World War and permanently on 22 September 1930. The rest of the branch – the GWR absorbed the local

Company in 1901, paying £6 per £10 share – has never been more than moderately busy: though continuing cheerful, Bridport has relatively lost in importance. Freight traffic has ceased and complete closure has been threatened for many years. As in East Devon, the absence of a main-line railway as well as natural features – the Chesil beach continues in diminishing extent from Weymouth past Abbotsbury to near West Bay – may be thanked for the preservation of coastal scenery.

CHAPTER X

The Somerset & Dorset

INDEPENDENT DAYS

Few railway centenaries have been celebrated with such gusto as the Somerset Central's at Burnham-on-Sea on 28 August 1954. The gaily decorated Bristol Channel resort offered a civic welcome to visitors, a procession, an exhibition, teas in the Town Hall and donkey rides for children. This astonishingly elaborate programme might have subjected Burnham to gentle ridicule even in the absence of two special facts which tickled the imagination of reporters. The first was that the line opened on 28 August 1854 didn't reach Burnham anyway, and the second that the extension later built to the town had been closed to passengers and removed from British Railways' maps several years previously.

Though enjoying the fun, visitors failed to understand the *raison d'être*. Not so the local people, who took for granted that the centenary of the opening of the first section of the Somerset Central should be well recorded – and at Burnham because it happened to be the most convenient place. The Somerset Central, of course, was the first and principal of the partners which joined to form the Somerset & Dorset. The S. and D! 'Slow and Dirty' or 'Swift and Delightful' – a railway which decidedly was never prosperous and never amazed its passengers by its efficiency but none the less provided an indispensable service and won the wholehearted love of all who knew it.

To live on the Somerset & Dorset was a romantic distinction. What made it so individual among railways: its cross-country character, the places it served and their melodious names, the straining of its locomotives when surmounting the Mendip ridge, the glorious and varied views from the train including that breath-taking glimpse from Masbury summit 811 ft above the sea, its almost comical poverty, the motley assortment of rolling stock? Whatever it was, many people otherwise not in the least interested in railways looked forward to using the Somerset & Dorset, though at best their trip would be miserably slow and at worst indescribable. Even the interminable manœuvrings at Templecombe were taken in surprisingly good part:

> In Wessex did you keep slow company
> With the "tanks" (royal) of the S. and D.,
> That waited, as it seemed, for crack of doom
> (While they performed strange rites) at Templecombe?

In the first edition in 1958 I went on to say: I have spoken as though all this is in the past. True, the outward glory of the line, the royal or prussian blue livery smartly lined in gold and black, is a mere memory; the Wells and Bridgwater branches, as well as the Burnham section, have been closed to passengers; and nationalization has involved dividing responsibility between Regions. But its individuality, if not unblemished, remains strong. Even dieselization would not bring complete rationalization. Such is its nature, cutting across the grain followed by the other railways of the West, that it must be treated separately even if it means going over some of the same country again.

By the time I revised the book for the third edition the *Pines Express* had ceased to follow its traditional route, through freight from the Midlands to the South West was no longer sent *via* Templecombe, and physically the neglected system still worked by steam locomotives not properly overhauled was in danger of collapsing. When closure had to be postponed because of delays in licensing replacement buses, it was only possible to continue with an emergency and much-reduced service. But the individuality remained till the end.

A whole book could be written about schemes to link the Bristol and English Channels to save the hazardous voyage round Land's End – so hazardous, that in a single year (1832) 100 ships on the passage were stranded along the coasts of the South West: 50 of them were seriously damaged or totally wrecked with heavy loss of life. The first canals to cut across the peninsula were proposed at the end of the eighteenth century. Railway schemes included lines from Padstow to Fowey, Bideford to Plymouth, Watchet to

Bridport and near Bridgwater to Lyme Regis. Despite such high-sounding titles as the Bristol & English Channels Direct Junction Railway, none was built – except the Somerset & Dorset, and that, ironically, did not begin with the aim of saving mariners' lives. As already stated, it began as two separate railways. The Somerset Central and the Dorset Central even had different gauges.

The Somerset Central, an independent Company fostered by the Bristol & Exeter Railway, was formed in 1852 and on 28 August 1854 opened a broad-gauge line from Highbridge Wharf to Glastonbury. Quaker businessmen had been prominent among the Glastonbury people wanting a line: they included James Clark, a shoe manufacturer. The present C. & J. Clark Ltd of Street were largely responsible for organizing the 1954 centenary celebrations, with some of the guests in period Quaker costume. At first, it should be noted, a line to Bridgwater had been favoured. but Highbridge had been chosen unanimously at a meeting of 'such fervour throughout, that it broke out at times in spirited ebullitions' held in June 1851. The advantage of the Highbridge route was that the line of the Glastonbury Canal owned by the Bristol & Exeter, could be used; the waterway was sold cheaply to the Company and closed. Soon after opening, the railway was worked by the Bristol & Exeter for a guaranteed 4 per cent on the capital, but the local Company retained Highbridge harbour and developed a substantial trade, including Caerphilly cheese destined for Wales but made (as it still is) mainly in Somerset.

A Government survey had recommended Highbridge as a potential packet port, but the Somerset Central had more ambitious ideas and on 3 May 1858 opened the extension to Burnham, where trains could run on to a 900-ft pier, costing £20,000, intended as base for a passenger business to Wales. The full story of the pier and the ships that used it has been told elsewhere, and later events concerning our story may be dismissed briefly here. The railway lacked power to run steamers, but a subsidiary Burnham Tidal Harbour & Railway Company was formed in 1860. At first the ship was secondhand, but a new one was ordered in 1866: owing to that year's financial crisis, she had to be registered as the property of her namesake, George Reed, 'the foremost citizen of Burnham', who was soon forced to sell her. Even when the Bristol Channel to English Channel route was finished, little pier

Page 201 Bygone days in Cornwall

Plate (45) Teetotal gala excursion crossing Redruth Viaduct in 1851.

Plate (46) The Bodmin & Wadebridge Railway's complete range of passenger rolling stock on a directors' special in 1886.

Plate (47) One of Brunel's last timber viaducts, Penryn, on the Falmouth branch in 1920; replaced by an embankment in 1923.

Page 202 Somerset & Dorset

Plate (48) Memory of hopeful days.

Plate (49) A 1956 Nottingham-to-Bournemouth express, with a 4–4–0 piloting a 2–8–0 example of the S. & D.'s goods class, about to enter Combe Down Tunnel near Bath.

Plate (50) Bridgwater North, the first station mentioned in this book.

Page 203 Dorset coast country

Plate (51) *A study of Weymouth station.*

Plate (52) *A 1912 down Somerset & Dorset express near Corfe Mullen on the Broadstone line with the old route to Wimborne running parallel.*

Plate (53) *A Swanage branch push-and-pull train.*

Page 204 A page from the scrapbook

Plate (54) Broad-gauge express passing Silverton, near Exeter.

Plate (55) On the level crossing at Poole at the turn of the century.

Plate (56) The first diesel multiple-unit train to visit the West at Totnes in 1958.

traffic resulted, despite the boast of the advertisement illustrated in Plate 48. Two years after the opening of the Severn Tunnel, the Somerset & Dorset abandoned passenger ships, but the Highbridge cargo service continued until 1933. Highbridge harbour closed in 1955.

It was the Somerset Central's avowed intention to join the Wilts, Somerset & Weymouth, and at first it was assumed this would be done *via* Wells and Frome. Powers for the Glastonbury–Wells section of the proposed route were obtained in 1855; but the Company was loath to use them, for almost immediately an alternative, less expensive scheme for a line from Glastonbury to near Bruton won the support of the majority of shareholders. As seen in Chapter II, it was only due to pressure from the minority that Wells was eventually given a dead-end branch, opened on 15 March 1859. The Bristol & Exeter assumed that even if the main line was continued southward to near Bruton the object remained to effect a junction with the broad-gauge Wilts, Somerset & Weymouth. But the 1861 Bill for the extension showed that there was no such intention. The Wilts, Somerset & Weymouth was cast aside, and narrow gauge asked for on the grounds that this was the gauge of the Dorset Central, which was to be joined. It was also proposed to lay a narrow rail on the Burnham–Glastonbury and Glastonbury–Wells lines. The Bristol & Exeter angrily gave notice to end the working of these and opposed the Bill so successfully that the continuation of the broad gauge to Bruton and the promised junction were insisted upon, though the extension of the narrow gauge to Burnham was allowed. Glancing ahead a moment, it proved a hollow victory for the Bristol & Exeter and the broad gauge cause generally, for though the earthworks were made the junction was probably not finished, and certainly never used. Hardly a broad-gauge train travelled on the Somerset Central as far south as Bruton, and about 1870 the broad-gauge rail was removed from the entire system.

The Dorset Central rose from the ruins of an 1852 scheme for a South Midland Union Railway, south from Mangotsfield roughly along the route later adopted by the *Pines Express*. At first only the 10¼ miles from the LSWR at Wimborne to Blandford were authorized, in 1856, but soon the Dorset Central was plotting to meet the Somerset Central and so establish coast-to-coast communication. Wimborne to Blandford, cheaply built with sharp curves and gradients, was opened on 1 November 1860, and temporarily worked by the LSWR. Next came the section from Templecombe to Cole, a place near Bruton, on 3 February 1862. The same day the Somerset Central opened its extension southward from Glastonbury to Cole. In both cases engineering works were light, the contour being followed wherever possible.

From the start, the lines on either side of Cole were run as a single unit, by the Somerset Central which provided all the rolling stock – narrow gauge, of course. The first of this should have been ready long before, in time to take over when the Bristol & Exeter's lease terminated. But it was not ready and even completion of laying the narrow gauge from Glastonbury to Burnham and Wells lagged. So nobly the Bristol & Exeter agreed to continue working the line on the broad gauge for a while. This helped foster the feeling among some shareholders that they had treated the broad-gauge Company most shabbily. One of them added that while it had always paid the 4 per cent guaranteed dividends promptly, there seemed little likelihood of payments on ordinary shares now that they had to stand on their own feet.

By an Act of 7 August 1862 the Somerset Central and Dorset Central amalgamated on 1 September 1862 under the title of Somerset & Dorset, and on 31 August the following year the Burnham–Wimborne route was completed by opening the missing 16 miles between Templecombe and Blandford. The LSWR allowed trains to reverse at Wimborne and continue to Poole, so the coast-to-coast service was a reality. Goods traffic immediately increased, tonnages at Highbridge doubling, but alas, nothing like enough patronage materialized to justify the amalgamated capital of £629,073. Indeed, we soon hear the last of the Somerset & Dorset as a coast-to-coast route, although in 1865 and 1866 the Company chartered *Albion* for a Poole–Cherbourg link 'in connection with the service between the North, Bristol, South Wales and Poole'.

The inclusion of Bristol in the list has a touch of pathos: the Somerset & Dorset dearly longed to extend there, but could not. Plans for a drive thence across the coalfield were produced each year – until the 1866 financial crisis. The last of a series of disruptions which had been occurring at about ten-yearly intervals, it nearly broke the Somerset & Dorset. Short of capital, the Company had borrowed from the revenue account to pay for part of the narrow-gauge rolling stock and other equipment. Now it was more difficult than ever to raise fresh capital – in fact impossible; there was no hope of recouping the revenue account and debenture interest could not be met. A receiver was appointed, and though seizure of property was averted, some engines ran only on creditors' suffrance, bearing an owner's plate to show they no longer belonged to the Company.

Schemes for capital rearrangement were rejected in 1868 and 1869, but in 1870 the Court of Chancery discharged the receiver and the Company was allowed to raise £160,000 in debentures. This brought the total consolidated capital up to £1,324,165, of which no less than £796,950 was in debentures. Most concerns would have tried to sell out long before this, but whatever else the Somerset & Dorset lacked – and there was plenty – it had courage and determination. It passionately desired not only to make the line pay – the optimists were incurable – but to serve the community by connecting places people might usefully travel between. No sooner had the directors retrieved control of their own affairs than a northward drive was again planned. It was too late to think of Bristol (Yatton–Cheddar was open and Bristol–Radstock looked like succeeding) and the opening of the Midland Railway's branch from Mangotsfield to Bath (to a temporary station in 1869 and the present one on 7 May 1870) made that city the obvious choice. The Bill for the extension there, from Evercreech, between Glastonbury and Templecombe, was passed in 1871 and confirmed two provisional arrangements: running powers over the Midland for the last half mile to Bath, and the purchase of the Somersetshire Coal Canal Tramway, of which six miles, between Radstock and Midford, were to be adapted for the route of the new line. The canal itself was mentioned in Chapter VIII; the tramway was in lieu of a 7¼-mile branch from Midford to Radstock and was opened in July 1815

> by which means a certain and regular communication is now opened to the several great collieries of Radstoke, Welton, Clandown and Smallcombe, from which works a constant supply of the very best Somersetshire coal may now be depended on to any extent.

Horses drew eight or nine waggons carrying 27 cwt. each, and according to the Somerset & Dorset's prospectus for the Bath Extension (as it was officially called) traffic was over 100,000 tons a year. The tramway's conversion from a canal feeder to a main-line railway was a bitter blow to the Kennet & Avon, which had most of the traffic handed over to it by the Somersetshire Coal Canal at Limpley Stoke.

Work on the Bath Extension began in 1872 and proceeded apace, with 3,000 men employed. Its 26 miles cost over £400,000 and involved extremely heavy works, including four tunnels and seven principal viaducts. Steep gradients to take the track across the Mendips included a gruelling start at 1 in 50 from Bath up to the second tunnel, dirty, noisy Combe Down, under Ralph Allen Park, and numerous sections also at 1 in 50 from Radstock to Masbury summit. The descent from Masbury to Evercreech was even sharper, with an almost continuous five miles at 1 in 50. The contrast between the new main line and the original one across the peat moors to Burnham could not have been greater. (Evercreech to Burnham soon became treated as a branch.)

Completion of the coast-to-coast route had once been expected to bring prosperity; now all hope lay in developing north-to-south traffic. Expectation ran high as opening day – 20 July 1874 – drew near. Shareholders looked forward to dividends at last, and as far away as Plymouth traders excitedly awaited the end of the broad-gauge monopoly of Midland to South West traffic. Only *Bradshaw* was sceptical. In the July issue it announced that the line was 'expected to open in a few days, when something like the following arrangements may take place'.

At first there were only four daily passenger trains – two carried Birmingham–Bournemouth coaches – but goods traffic developed dramatically. Between 1 October 1874 and 31 August 1875, we are told, 107,529 tons were exchanged with the Midland Railway at Bath and 89,431 with the LSWR at Templecombe. (Exchanges at Highbridge and Wells were 12,166 and 5,436 tons.) It is dangerous to give a starving man too much food at once: the Somerset & Dorset could not stand the strain. Engines, track and staff quickly showed signs of wear. Each day brought increased anxiety about the maintenance of the service – but that was nothing compared with the general financial worry. Such was the

Company's plight that Bath Extension shares had been sold at discounts of up to 50 per cent. Profits even now were inadequate to provide proper maintenance or pay all debenture holders. As no further scheme of rearrangement could have been acceptable, the only course was to end independence.

The GWR was approached first. It brought the Bristol & Exeter into the picture and after leisurely negotiations suggested to the LSWR that it might like to take that part of the system south of Templecombe. Scott, the LSWR's manager, was horrified to learn that the Somerset & Dorset, vital to the narrow-gauge cause in the West, had offered itself to the GWR and determined to stop at nothing to prevent the sale. While his Board was supposed to be considering the south of Templecombe offer, he rushed to Birmingham to meet Allport, manager of the Midland Railway. Consultations were followed by a hurried vetting of accounts and a joint offer to the Somerset & Dorset more favourable than that provisionally suggested by the GWR and Bristol & Exeter. The offer was accepted while the latter Companies were still waiting for the LSWR's reply. Perhaps the word 'still' is ill-chosen, for the complete circle of negotiations took place during one hectic week-end of August 1875. Daniel Gooch, the GWR's chairman, accused the LSWR of breach of faith and a lengthy public correspondence ensued. It scarcely excuses the tactics used by Waterloo to add that the GWR was not the natural or most desirable successor to the Somerset & Dorset, and almost certainly would not have stimulated trade on the scale of the Midland Railway and LSWR.

These were placed on an equal footing in all respects by an Act of 13 July 1876 which confirmed the 999-year lease, to start from 1 November 1875, rental being fixed at £43,056 for the first year, rising to £57,408 in the fourth year, plus one-tenth of any sum by which gross receipts exceeded £114,816. The Somerset & Dorset was thus able to pay all debenture dividends and 3½ per cent on Bath Extension ordinary stock. The total capital actually spent before the lease became effective was £2,108,503, plus £223,526 on rolling stock. Out of the latter sum a locomotive stock of only 26 had been provided by 1876. The first narrow-gauge engines – 2-4-0s – had arrived in 1861. Almost immediately the story becomes complicated, for the Company ordered more than it could actually buy. Later, some of the light locomotives had to be sold in order to buy a few better able to climb the Bath Extension grades. Livery was dark green, with one exception – *Bluebottle*, a 2-4-0 tank which retained the royal blue in which it had been exhibited by its makers at Hyde Park in 1862. It was transferred to the LSWR in 1871, but the stir it had caused meanwhile among Somerset people is said to have influenced the later decision to paint all Somerset & Dorset stock blue.

FROM LEASE TO NATIONALIZATION

If the Midland Railway and LSWR did not appreciate the seriousness of the liability they had taken on, it was forcibly brought home only a few weeks after Parliament had confirmed the lease. On 7 August 1876 thirteen people were killed in a head-on crash near Radstock. Despite the deaths, it was the cause, not the result, of the accident which shocked the public. One scarcely knows where to begin describing the line's 'positive state of disorganization' that Bank Holiday. Of the three signal boxes involved, two (Wellow and Radstock) controlled crossing places and the third (Foxcote) merely divided the single line between the others into two sections. This in itself was illegal, as Board of Trade regulations stipulated that each section between passing places should be treated as one unit only. The signalman in the Foxcote box was a boy who 'could not write or read excellent' and lacked strength to pull the distant signal lever properly. Making allowances for this, drivers assumed the line was clear even if the arm was 'nearly at caution'. That evening the signal lamps were not lit as there was no oil. At Wellow and Radstock conditions were only slightly better. The entire staff was suffering from fatigue, having been on duty well over twelve hours in most cases.

Anxiety was first felt when a Wimborne–Bath relief train was literally lost by the 'crossing agent', a functionary who, tucked away in his office at Glastonbury, controlled traffic in virtually hit-or-miss fashion. Having failed to locate the train elsewhere, he called Radstock. The signalman there had been annoyed by the crossing agent on a previous occasion and took the opportunity for revenge. He answered the first question about the train's whereabouts by 'Don't know', the second and third questions by 'Good' and the fourth – in answer to 'Do you refuse to answer my question?' – by another 'Don't know'. To crown it all, when a down empty-stock train left Wellow, the fact was unrecorded on the block instrument. The signalman at Foxcote knew nothing of it, and shortly afterwards offered the Wimborne–Bath train to Wellow; the signalman there accepted it unhesitatingly. The crossing agent was told of the accident at 11.37 p.m. and did not leave his post until he had been on duty twenty-four hours.

Following this, Edward Tyer, a great name in signalling history, met public demand for greater safety on single lines everywhere by inventing the electric tablet instrument, still used in its various forms on most single lines today. Tablets, tokens or staffs are interchangeable between instruments at either end of the section. Only one from the combined stock of tokens may be withdrawn at one time, and that, of course, goes with the engine driver. The disadvantage is that a token has to be set down and another picked up at each signal box. Exchange is usually by hand, involving speed restriction, but the Somerset & Dorset pioneered automatic exchangers, the principle being similar to that used in the exchange of mail bags at speed by travelling post-offices.

But better signalling arrangements were a small point in the improvement campaign of the leasing Companies. Financial rearrangements soon enabled them to take control and replace the old Board by the new Joint Committee. As an incidental to this, the general manager since Somerset Central days, Robert Arthur Read, a man of infinite energy and drive, sadly ended his association with the line he had done more than any other single person to create.

The reversal at Wimborne was abolished for Bournemouth expresses by the opening of the direct Corfe Mullen–Broadstone line on 14 December 1885. (Gradually local trains also took the new route, Corfe Mullen–Wimborne being closed to passengers on 12 July 1920 and used by the last goods on 17 June 1933: the first half-mile at Corfe Mullen has been retained as a clay works siding.) Traffic was increasing fast – it doubled, for instance, between 1876 and 1881 – so part of the Bath–Broadstone main line was doubled. Wincanton to Templecombe was doubled in 1887, Midford to Wincanton followed in stages up to 1894, and Blandford to Corfe Mullen followed in 1901. That left single Bath to

Midford through Combe Down Tunnel, Templecombe to Blandford, and Corfe Mullen to Broadstone – 23 out of $63\frac{1}{4}$ miles, as it remained. Where viaducts were doubled, a second was attached to the first. (On 8 February 1946 two of the 27 arches of the Charlton Road Viaduct, Shepton Mallet, collapsed and single-line working was necessary until replacement.)

Rolling stock was improved beyond recognition. The Midland Railway took chief responsibility for the locomotive side, quickly bringing in new machines and relegating most of those already on the line to local and branch work. The locomotives were painted in the Midland's pale green livery until 1883, when the Somerset & Dorset as well as the Midland flared red for both engines and coaches. So royal blue, introduced as a Somerset & Dorset speciality soon afterwards, was the fourth locomotive livery on the line. Four-coupled tanks worked even the expresses until 1891, when some tender 4-4-os, similar to the Midland's own engines of that wheel arrangement, went into service. From now on nearly all engines – including several series of 0-6-0 goods – were adaptations of Midland classes. The only significant exceptions were 2-8-0 tender goods engines, which with a tractive effort of 36,000 lb were among the largest and most powerful in Britain. They were naturally used more particularly on the Bath–Evercreech section. Only three tank engines, but a fairly wide range of coaches, were built at Highbridge works.

The first genuine express appeared in 1880. From then until 1914 came a steadily widening selection of through services, including the all-the-year-round Manchester–Bournemouth restaurant car train, begun on 1 October 1910. It was not at first called the *Pines Express*. The profit margin was now reasonable, but after the war much local and medium-distance traffic turned to the roads, and revenue failed to keep pace with rising costs. In 1910 income was £235,686 and expenditure £179,431; in 1929, for instance, income was £375,794, but expenditure (of course not counting dividends on capital) £387,403. In both years goods brought in more than passengers. To the delight of enthusiasts, the 1923 grouping did not spoil the line's unique character, the LMSR and Southern merely becoming joint owners replacing the Midland Railway and LSWR. But mounting losses made economies essential, and in 1930 responsibility for various functions was assumed directly by the big Companies. The Southern Railway took on the track and signalling; the LMSR traffic organization, all locomotives (now painted in its standard livery) and most rolling stock. The staff of just over 500 was considerably reduced and Highbridge works closed.

Even so, the system continued to make a loss. The main line was inevitably expensive to work, with high seasonal variations of traffic. For most of the year this did not justify the second track on the doubled sections, but at summer weekends long delays occurred in crossing trains on the remaining single sections. On summer Saturdays until 1962 a continuous stream of long-distance expresses ran to Bournemouth from starting-points such as Manchester, Bradford, Nottingham and Cleethorpes. Even on normal days average speeds were low, the *Pines Express* being allowed 2 h. 27 m. from Bath to Bournemouth (71½ miles), of which 57 m. were consumed on the first 26½ miles to Evercreech Junction – though occasionally the Junction was reached so early that the tank engine assisting perhaps a 'West Country' class locomotive thus far was allowed to run to Templecombe ahead of the main train. It was always one of the joys and sorrows of the Somerset & Dorset that even the wind blowing over the Mendips could profoundly affect speed.

Then, in September 1962, the *Pines Express* was re-routed *via* Reading and Basingstoke, leaving the S&D like Hamlet without the ghost. Later the through freight was diverted, and even football excursions between Bath and Bournemouth were sent another way. Steadily the system ran down, some of its stations and signal-boxes almost literally falling to pieces. Closure seemed imminent for several years and eventually took place on 7 March 1966 after a two-month reprieve during which only an emergency service had operated. After nationalization the system became a pawn in inter-Regional adjustments: at one stage three Regions had a hand in its affairs; now most of it is deeply embedded in Western territory. Of the effect of the changes on the line's efficiency, the less said the better. But diversity of management added to locomotive interest. Somerset & Dorset survivals, Southern (including 'West Country' Pacifics), Great Western and British Railways standard locomotives were all to be seen in the 1950s. The line was steam-operated to the end, and standard engine headcodes still did not apply, passenger trains being denoted, as always, by lamps under the chimney and over the right buffer.

The Somerset & Dorset's economic influence has of course been greatest on places outside its own system, notably Bournemouth. The completion of the first narrow-gauge route from the Midlands to the South West by the opening of the Bath Extension in 1874 helped Plymouth in particular. Of lesser importance, it spelt doom for many small breweries in the South West, Burton beer being one of the chief commodities carried. Other beer came from the famous Oakhill brewery, connected for some years after 1904 by a two-mile narrow-gauge tramway from Binegar station. Brewing has also been important at Shepton Mallet. Here, as at Radstock, the arrival of the Somerset & Dorset, even late in railway history, was a fillip to trade. Shepton Mallet has long depended more on the newcomer than on its first railway, the branch from Witham (see Chapter II), though it was this which enabled the town to foster new industries, notably silk and crêpe, after the failure of the woollen trade which had once employed 4,000 people.

The only sizeable town served exclusively by the Somerset & Dorset main line was Blandford; the route from Broadstone as far as here was retained for goods traffic until 2 August 1965. A Blandford Camp Railway, two miles long, opened late in 1918, closed at the end of 1919 and reopened for six months in 1920–1. Surprisingly little agricultural produce was carried on the Somerset & Dorset, though some milk came from a creamery at Bailey Gate. Of the traffic originating on the system, coal was always by far the heaviest, ever since 1874. Up to 200,000 tons a year were once despatched by the S & D (mainly to the south) from the Radstock area. The decline of the North Somerset coal field has already been mentioned (pp. 178–9). One of the major collieries, Norton Hill, closed almost simultaneously with the railway. The sole surviving S & D colliery, Writhlington, is now served by a connection to the GWR's North Somerset line at Radstock and this is now the only piece of S & D track still in use.

Use by local passengers was greater to the north of Templecombe than to the south: to some extent the line acted as a feeder to London services, *via* Templecombe. But if Bath had one place where London's very existence could be forgotten, it was Green

Park station, a terminus with two platforms, two centre roads, an all-over roof and an impressive classical frontage, which wore a decayed air and in winter at least was nearly always pervaded by Sabbath calm. The building still more or less stands, horribly neglected, the station being used as a car park until its future is determined.

TEMPLECOMBE AND BRANCH LINES

Arrangements were always diabolical at TEMPLECOMBE, where the Somerset & Dorset was crossed at right angles by

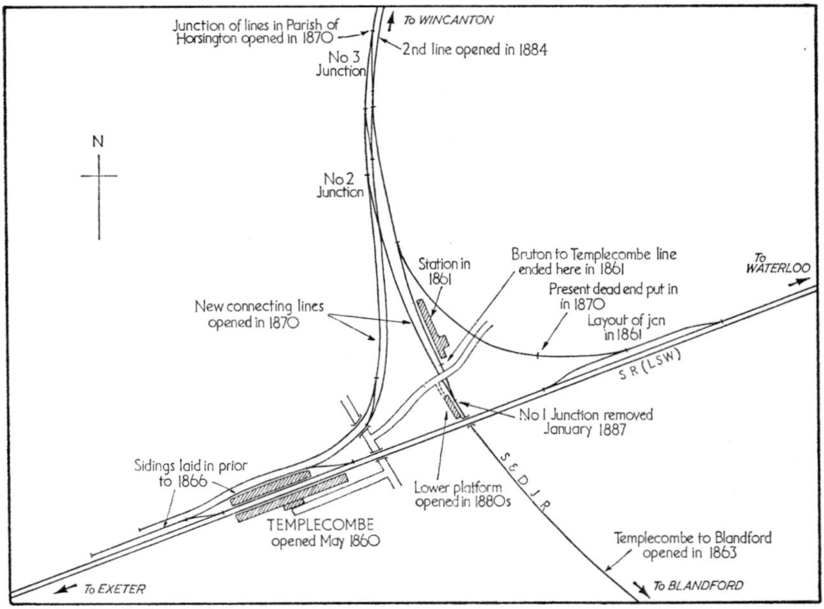

the trunk Waterloo–Exeter route. Originally the connection between the two lines was to permit through working from Burnham towards London. Except for a short spell when the LSWR provided a shuttle service between the two stations concerned, the arrangement was that most trains from Burnham backed down the up Exeter–London track to reach high-level Templecombe Junction. Some called in addition at the Somerset & Dorset's own low-level station. But in 1870 the Salisbury & Yeovil Railway built its only branch, of 33 chains from Templecombe Junction (on the Exeter side of the intersection of the two routes) towards Burnham. A separate platform was provided at Junction for Somerset & Dorset trains. The original connection was closed as, later (1887), was the low-level station – though at about the same time a new Templecombe Lower Platform was opened for a few trains not calling at Junction. It remained the terminus for a Saturday evening local until 1965. A few expresses omitted Templecombe; the other trains from Bath to Bournemouth were bound to back out of Junction – now called just Templecombe – before resuming their southward journey, while northward trains had to back into the station. In each case there were locomotives at both ends.

When first relegated to a branch, Evercreech Junction to BURNHAM remained fairly busy, but this century has brought continual decline, though parcels and goods traffic remained heavy at Glastonbury and peat was handled at several of the remote stations on the section through the willow and peat farms, some with their own narrow-gauge systems, heaths and ditches, all only a few feet above sea level. Highbridge station, with its four platforms and extensive layout, was amusingly large for later needs; just beyond was the level crossing over the main Western line. It was the difficulty of preventing delays to Western trains while Burnham ones were using the crossing which prompted the closing to passengers of Highbridge to Burnham on 29 October 1951. Until then this section was busier than the rest of the line and Burnham's development as a resort owes much to it. Today passengers make their way by road from what is called 'Highbridge for Burnham-on-Sea' station. But until 1962 an excursion train still ran to Burnham proper twice a week in summer. Closure to goods at Burnham was on 20 May 1963. The Evercreech–Highbridge section remained open to passengers until the closure of the S & D main line on 7 March 1966; Highbridge–Bason Bridge was used for milk traffic until October 1972 when the site was absorbed into works connected with the M5. But there remains ample to attract the industrial archaeologist in the route of the S & D for many years to come. Every time my Taunton–Bristol express rushes through Highbridge I glance at the remains of the works which were once the pride of the area – and almost painfully miss the clatter of crossing the S & D line at an acute angle.

The WELLS branch, beside the Evercreech Junction line for the first mile from Glastonbury, was described in Chapter II.

The BRIDGWATER line, mentioned at the beginning of the book, was worked as part of the Somerset & Dorset, though leased by the Bridgwater Railway (to be strictly correct, Bridgewater Railway) to the LSWR only. For some years after the opening on 21 July 1890 semi-fast trains were run to Templecombe to catch sharp connections for London. Optimistically, the bridges were built for double track; but the single track was closed to passengers on 1 December 1952 and completely on 1 October 1954. A wharf branch in Bridgwater was closed during the First World War and its rails removed in the second.

CHAPTER XI

On the Rails

RISE AND FALL OF THE BROAD GAUGE

Though the whole of the West Country has now been covered, a further chapter is needed for several aspects which overlap a number of chapter areas, notably the development of through services to the South West.

Brunel invented the broad gauge for speed, but little was seen in the early days of the GWR proper and it was not until the opening day of the final section of the Bristol & Exeter Railway into Exeter that the first indication of what could be achieved in the way of prolonged fast running occurred. As already noted (and see Appendix III), Gooch drove the return directors' special to Paddington, 193¾ miles, in 4 h. 42 m., including stops. But in January next year (1845) 7 h. 10 m. was thought good enough for the best regular train, the night mail. During the spring, however, a Government inquiry was held into the gauge question, and to demonstrate that the broad possessed advantages to offset the growing inconvenience caused by lack of uniformity, on 10 March a daytime train was accelerated to reach Exeter in 5 h., which – to the amazement of the world, for no other railway at home or abroad offered anything comparable – was cut to 4½ h. from 12 May.

But broad-gauge initiative was fickle, and after 1852 the wheels gradually slowed down, first of all on the Bristol & Exeter section. In 1848 the chief express was extended over the South Devon Railway, reaching the temporary station, at Laira, near Plymouth, in 6 h. 50 m.: after 1852 the time to Plymouth proper (Millbay) grew to 7 h. 35 m. In 1849, incidentally, the express was nicknamed the *Flying Dutchman* after the racehorse which won both the St Leger and the Derby that year.

This train, of course, was exclusively first and second class. Until 1853 third-class passengers were unable to travel from London to Plymouth in one day. Then, by the introduction of what was colloquially known as the 'Plymouth Cheap', they were able to leave Paddington at 6.15 a.m. and spend a mere 12 h. 20 m. *en route*, being overtaken at Exeter by the express, which had started 2½ h. later and was itself scarcely rapid. Despite all discouragements, however, the Plymouth Cheap was popular and must have paid its way handsomely. Passengers included those preparing to sail from Plymouth to the New World and, on the upward journey, West Country boys and girls off to London to seek their fortunes. For better or worse, the lives of thousands of families were altered by the Plymouth Cheap.

During the first two years of its trunk route to Exeter, the London & South Western made few bids for popularity, but in February 1862 it suddenly decided to reach the city in 4¾ h., or 25 m. less than then taken by the best broad-gauge train. Its route of 171½ miles was considerably shorter, but its gradients much the heavier. Rubbing their eyes, on 1 March the broad Companies reintroduced a 4½ h. timing with an 11.45 a.m. from Paddington, which officially adopted the dormant title of *Flying Dutchman*. At first the South Devon took it on to Torquay only, but two years later it went to Plymouth as that Company's first real express, covering the 53 miles from Exeter in 1¾ h. Plymouth was thus brought within 6 h. 25 m. of London for first- and second-class passengers willing to pay the excess express fares – between London and Exeter, for instance, 5s on the ordinary 35s or 25s respectively.

But again enthusiasm collapsed. Services to Exeter by both routes were steadily slowed, and at the end of 1867 the *Dutchman* disappeared altogether for eighteen months. When reinstated it took 6¾ h. to reach Plymouth; the best up train took no less than 7 h. 35 min. In May 1871, however, the times were cut to 4¼ h. to Exeter and 6¼ to Plymouth, at which they remained for twenty years.

Meanwhile third-class passengers had been admitted to a secondary express in each direction, not of the GWR's free will but of necessity to meet LSWR competition at Exeter. To the august GWR, Waterloo seemed ever jacobinic and entirely lacking in decorum. It positively encouraged third-class business! How the Paddington hierarchy hated the changes in the social order effected largely by the railways themselves. This crass conservatism was perhaps best demonstrated by its attitude to express fares. These were abolished on the Bristol & Exeter and South Devon in 1871, and on the LSWR in 1877, after which the GWR charged them only for journeys where there was no alternative LSWR route. This naturally evoked outcries about the dangers inherent in monopolies.

Not only did the excess fares and the restrictions involved in the issue of ordinary first and second tickets cause untold complications (to add to the confusion, third-class fares now cost 1¼d a mile, though Parliamentary trains at 1d a mile remained a legal obligation), but the Press lost no opportunity to point out that GWR charges were in any case higher than average: 2½d and 1¾d per mile ordinary first and second class, compared with 2d and 1½d on most trunk lines. As already mentioned, the Bristol & Exeter extracted even more, and the difference remained on that part of the line even after it became part of the GWR. Worse, express fares were then reapplied throughout such journeys as Paddington to Taunton. During the last five years of these fares, the only other companies charging them were those of South East England, poverty-stricken and with none of the GWR's advantages.

The lack of business acumen was coupled with stagnation on the mechanical front, largely inevitable as the doom of the broad gauge was already sealed. For its part the LSWR was keen enough to drive new wedges into broad territory but had scant inclination to improve services. When it reached Plymouth in 1876, for example, traffic was gained less through its own efficiency (the best of the four daily trains took all but seven hours) than through the greater lack of enterprise of the broad gauge. The LSWR's seven-hour train left Waterloo in the afternoon, and not for three years would Paddington counter with a similar one. However, when the 3.0 Paddington to Plymouth did start – it was known as the *Zulu*, from the war then being fought – it was five minutes quicker even than the *Dutchman*.

A special class, the powerful Plymouth tank engines, had been designed for the gruelling LSWR Exeter–Okehampton–Plymouth work, but owing to inadequate fuel and water storage were soon replaced by the ordinary tanks, beautifully proportioned and smartly finished in a deep chocolate brown, of Joseph Beattie, the first of several notable locomotive superintendents to come the Company's way. Even in engine design there was a sharp contrast between the GWR, which retained the same basic features of profile throughout most of its existence, and the LSWR, with individualists such as William Adams and Dugald Drummond as well as Beattie at the helm. Their experiments produced several first-rate designs, and some dismal failures with resultant modifying and rebuilding. This makes the LSWR's locomotive story infinitely too complicated for this book, which of course does not purport to be a mechanical history – though one class deserves special mention, Drummond's large 4-4-0 T9s, the 'Greyhounds', familiar on West Country lines for over half a century. Built in 1899–1901, they represented the prime of the LSWR: all lasted to be included in British Railways' stock list after nationalization. But in the 1870s and 1880s, our present concern, no tender express engines found their way to Plymouth. Even the *Flying Dutchman* trundled over the last part of its journey behind an impudent 4-4-0 South Devon saddle tank. Coaches were gradually improving. By the end of the 1880s the LSWR's best expresses were formed of neat bogie stock instead of uncouth strings of four-wheelers; and though few GWR locomotives were seen west of Bristol even after 1876, their smarter carriages certainly were. The locomotives followed just in time to become familiar objects on the road to Plymouth before the broad gauge died.

Indeed, just before its death it rallied remarkably. In June 1890 a new express appeared; it reached Bristol in 2½ h. (a speed not equalled since forty-two years previously!), Exeter in 4 h. 5 m., Plymouth in 5 h. 35 m. and Penzance in 8 h. 42 m. The Plymouth stop was North Road, the express being the first to by-pass Millbay. It was in fact the first train to remain something like an express all the way to Penzance. But the *Cornishman*'s most remarkable feature was that it carried third-class passengers. The end of the rigid régime had come: till October 1890 the old *Dutchman* valiantly struggled to keep select, but then it too bowed down. Every train on the line was open to the working-class. The Midland Railway had achieved this eighteen years before!

Whatever its limitations, the GWR had one virtue: majesty. The sentiment is not original, but deserves repetition. The old world stateliness, the spaciousness and solidarity of the *Cornishman* in its two broad-gauge seasons was never forgotten by those who saw it. The clerestory coaches were built to dimensions impossible anywhere else. The first-class compartments, 10 × 7 ft and resplendently furnished, were expected to contain no more than eight passengers at most. Of the GWR broad engines – what names they had – Sir W. M. Acworth wrote:

> No traveller on the line will see without a pang the stately *Iron Duke*, the wandering *Tartar*, or the swift-flying *Swallow* disappear from the road that has known them for 40 years. No engines in the world have so long and so famous a history as these old engines of Sir Daniel Gooch. Save that they have lost their sentry box at the back of the tender, from which the guard used to keep watch to see that his train was duly following, they look today, with their great 8ft driving wheels and the old-world brass-mounted boilers and brass wheel covers, just as they must have looked 40 years ago, when our fathers gaped open-mouthed at the tale of their achievement.

Gooch, eventually chairman of the railway he had served since before his twenty-first birthday, knew that the end of the 7-ft way was inevitable, but his own death relieved him of having to sign the death warrant. It was decided to convert the remaining broad lines to narrow between the night of Friday 20 May 1892 and the following Monday morning. The table on page 223, taken from MacDermot's *History*, shows what little was eventually left of the kingdom.

The mixed lines, of course, presented no trouble, the broad rails being removed at convenience. But during the one week-end it was necessary to change every inch of the 213 miles west of Exeter still exclusively broad. Moreover, only 36 of these miles had been modernized with cross sleepers. The rest of the track was, as from the start of the broad gauge, on longitudinal timbers. The longitudinal beam under one of the rails as well as the rail itself had to be removed to the new position. Preparations were made with the utmost thoroughness. Bolts and screws were slackened, sidings narrowed in advance, narrow locomotives brought down on broad

MILEAGE

Broad and mixed gauge lines in May 1892

Section	*Broad* *Double* M.	C.	*Broad* *Single* M.	C.	*Mixed* *Double* M.	C.	*Mixed* *Single* M.	C.
Main line	M.	C.	M.	C.	M.	C.	M.	C.
Paddington–Exeter					194	7		
Exeter–City Basin Junction[1]			1	25			1	25
City Basin Junction–Dawlish	10	58						
Dawlish–Parson's Tunnel			1	32				
Parson's Tunnel–Rattery	19	73						
Rattery–Hemerdon			11	20				
Hemerdon–Tavistock Junction	4	34						
Tavistock Junction–Millbay Docks Junction					3	45		
Plymouth, Millbay Station		10						
North Road–Devonport Junction (Loop)						24		
Cornwall Junction–Devonport					1	5		
Devonport–Keyham Junction							1	36
Keyham Junction–Truro			50	78				
Truro–Penzance							25	69
	35	15	64	75	199	1	28	50
Branches								
Swindon, on Gloucester line						60		
Bristol Harbour							1	24
Dunball Wharf								36
Bridgwater Docks								76
Exeter City Basin								34
Newton Abbot–Moretonhampstead			12	13				
Newton Abbot–Torquay	5	68						
Torquay–Kingswear			8	66				
Churston–Brixham			2	6				
Ashburton Junction–Ashburton			9	20				
Totnes Quay				62				
Tavistock Junction–Marsh Mills						18		
Marsh Mills–Lydford							19	15
Lydford–Launceston			12	34				
Laira Junction–Friary Junction						59		
Friary Junction–Sutton Harbour	1	3						
North Quay Wharves				32				
Burngullow–Drinnick Mill			3	10				
Truro–Falmouth			11	68				
Hayle Wharves								55
St. Erth–Lelant							1	8
Lelant–St. Ives			3	14				
	6	71	64	5	1	57	24	8
Summary								
Main line	35	15	64	75	199	1	28	50
Branches	6	71	64	5	1	57	24	8
	42	6	129	—	200	58	52	58

	Broad M.	C.	Mixed M.	C.
Total Length	171	6	253	36
Total of Single miles	213	12	455	14

[1] Down line only was mixed gauge.

trucks, and 4,200 platelayers assembled. Drawn from all parts of the GWR, many slept in tents and station buildings and as at earlier and lesser conversions drank an oatmeal mixture supplied by the Company. Stronger drinks were forbidden, and in Cornwall the engineering feat was claimed as a temperance victory.

The broad gauge was essentially a West Country institution

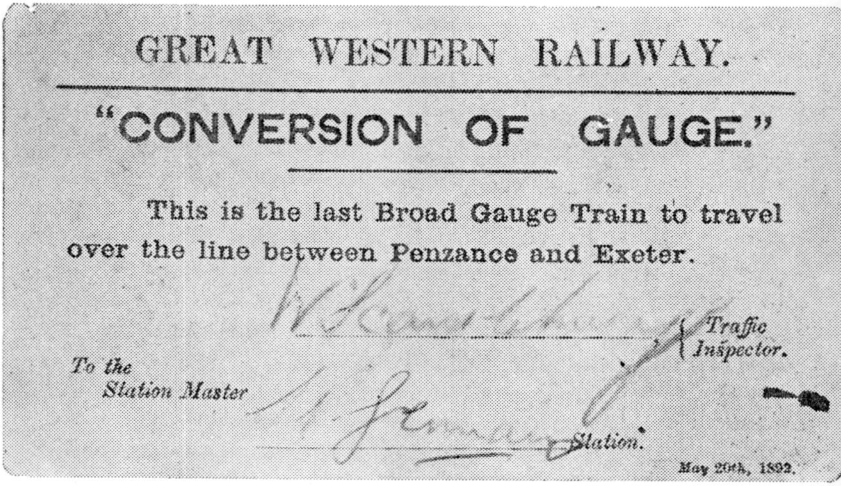

GREAT WESTERN RAILWAY.

"CONVERSION OF GAUGE."

This is the last Broad Gauge Train to travel over the line between Penzance and Exeter.

Traffic Inspector.

To the Station Master

Station.

May 20th, 1892.

and, past lethargy forgiven, on its last day the local population turned out in force with demonstrations of affection and loyalty. At Exeter the last down train was delayed while a gentleman standing by *Iron Duke* lectured on bygone glories; at Dawlish as it drew alongside its opposite number, passengers spontaneously leaned out of the windows to join hands in 'Auld Lang Syne'. Varied were the epitaphs chalked on the carriages. The Ashburton branch engine went to meet its end at Swindon draped in black crêpe presented by the Portreeve. That was on Friday night. As soon as the last empty train conveying the necessary certificate had passed, the executioners went to work, in gangs of 20. Thanks to their enthusiasm and perfect weather, narrow locomotives began testing parts of the line before dusk on the Saturday, and the down night mail, which had travelled between Exeter and

Plymouth *via* the LSWR, continued from Plymouth into Cornwall at 4.40 a.m. on Monday morning, after which the full service was restored. As the *Torquay Times* put it: 'Comfort has had to give way to convenience.' The announcement of a through express from Liverpool was 'pregnant with great possibilities for the future of Torquay, as well as of other Devon and Cornwall watering places. We have reason then to be thankful to the Great Western Railway Company for having relieved us from that ostracism for which the exceptional gauge was responsible.'

THE OCEAN TRAFFIC

A great future had been forecast for Millbay Docks but Plymouth, like Falmouth and Dartmouth, was to watch most of the ocean traffic go to Liverpool and Southampton. As the century wore on, however, a growing proportion of homeward-bound ships, especially from South Africa and India, stopped to drop their mails. The GWR took these off by tender and rushed them by special from Millbay to London, while passengers were left on board for the LSWR at Southampton: but this gentleman's agreement ceased after 1892.

At first the passenger specials to Paddington were not particularly fast, though 4 h. 27 m. for the 246¾ miles *via* Bristol once attained in 1893 suggested possibilities. It was early in 1904, when the LSWR also set out to attract ocean passengers at Plymouth, that the fun began. By then most of the GWR's route through South Devon had been doubled and of course for fourteen years the LSWR had possessed independent double-line access to the Three Towns by the Plymouth, Devonport & South Western Junction. The LSWR's ocean terminal was at the end of the Stonehouse Pool branch, hitherto used only for goods. While finishing touches were being put to it, the GWR consistently speeded up its specials. The LSWR challenge came on 23 April, when its luxurious corridor train flashed by the 230 miles from Stonehouse Junction to Waterloo in 4 h. 3 m. The GWR's famous reply, albeit with a mail special only, not carrying passengers, was on 9 May. The North German Lloyd *Kronprinz Wilhelm* left New York at 3.10 p.m. on 3 May and anchored off Plymouth at 8.0 a.m. on the 9th. Mails and bullion – American payment to

France for the Panama Canal – were hurriedly transferred from ship to tender and tender to the train of five big eight-wheeled vans, and *City of Truro* set off on the historic run recorded in the *Railway Magazine* of the day:

Millbay Dock Crossing	dep.	9.23.10
Plymouth (North Road)	pass	9.26.17
Newton Abbot	"	9.59.52
Exeter	"	10.22.12
Whiteball Summit	"	10.41.41
Taunton	"	10.50. 1
Bristol (Pylle Hill avoiding line)	arr.	11.26.29
	dep.	11.30.12
Bath	pass	11.43.50
Chippenham	"	11.56. 0
Swindon	"	12. 9.49
Didcot	"	12.29.20
Reading	"	12.42.21
Slough	"	12.55.15
Paddington (platform end)	arr.	1. 9.30
" (final stop)	"	1. 9.58

The average speed, including the Bristol stop, was 66·5 m.p.h.; between Exeter and Bristol 70·2, and between Bristol and Paddington (*City of Truro* was replaced by a 7 ft 8 in. single, and the load was only four vans) 71·5 m.p.h. In descending Wellington Bank a maximum speed of 102·3 m.p.h. was recorded. Some experts maintain that a 'City' was incapable of such pace, but if doubt exists the claim has not been satisfactorily disproved.

The LSWR could not better this: already there had been anxiety about excessive speed through Salisbury. But the emphasis remained on timing – the LSWR bitterly complained about the GWR's insistence on stopping all 'foreign' trains at Exeter St David's – until early on 1 July 1906 a special hauled by an almost new Drummond 4-4-0, large-boilered and with 6 ft 7 in. driving wheels (as was customary, it had replaced the Plymouth engine at Templecombe) reeled on the curve actually in Salisbury station and struck a train of down milk empties. The flimsy, if luxurious, wooden coaches disintegrated in fearful fashion, and 24 of the 43 passengers who were being rushed to London in time for a sleep before a late breakfast were killed. It was a sad day for the LSWR: on this very 1 July the GWR opened its shorter Castle Cary route to the West.

What the LSWR could not achieve in speed it tried to make good in comfort, and some fine sleeping-cars were introduced in 1908. 56 ft long and 9 ft wide, they each contained seven single and two double berths; but 'berths' is hardly the word: they were brass bedsteads. Heavy upholstery was avoided, simplicity for the sake of hygiene at last being taken into account. But the cars were not popular, and in 1910 were sold to the GWR. For a long time hostility between the two Companies had been growing. In 1907, for example, the GWR introduced a Plymouth–Brest boat service to which the LSWR replied with a Southampton–Cork service. But 1910 brought negotiations, and next year the Companies signed a publicly-acclaimed peace treaty which proved lasting. The LSWR ended all boat activities at Plymouth, and the GWR withdrew the Brest, and shorter-lived Nantes, services. Plymouth to Paddington traffic – there has seldom been much in the other direction – fluctuated considerably thereafter. It reached a peak of 45,300 passengers in 1930, but built up slowly after the war. In 1952 new accommodation for ocean passengers was opened at Millbay, and in 1957, for example, 19,203 passengers and 40,739 bags of mail were landed. Ocean specials remained a familiar feature until November 1961, when the French Line ceased to use the port, and one of the two tenders was withdrawn. Restrictions on the immigration of West Indians also reduced the traffic: tens of thousands of West Indians first stepped on to British soil at Plymouth during the 1950s. Complete boat trains had virtually ceased by 1965, although for another year or so passengers were still taken by rail from the dockside in coaches added to ordinary trains. Now all rails even for freight have been removed from Millbay.

A number of luxury saloons were introduced on the boat trains in 1929 and in 1931 came eight named after members of the Royal Family – the finest rolling stock Swindon ever produced. An example survives on the Dart Valley Railway. These, no longer named, have been used on some extremely fast runs even since the war, including the record 3 h. 37 m. for the 227½ miles from the Docks to Paddington on 7 October 1954. Appropriately enough the 'Castle' locomotive was *Isambard Kingdom Brunel.*

THE GREAT WESTERN BECOMES GREAT

It was not, of course, only the boat traffic which gained from gauge standardization. The first summer, 1892, brought the forerunner of the *Torbay Express*, as well as the Liverpool train already mentioned. Schedules for that season were perhaps theoretic, speed being limited over the rebuilt track, but in 1893 you could leave Liverpool at 10.40 a.m. and be in Torquay at 6.31 p.m. The *Cornishman*, the first corridor lavatory train the West had seen, left Paddington at 10.15 and after stops at Swindon, Bristol and Exeter reached Plymouth North Road 5 h. 38 m. later. In the summer of 1896 a regular advance Newquay section of it achieved the world's longest non-stop run – to Exeter, in 3¾ h. The first non-stop run between Plymouth and London was in March 1902, by a special which was returning King Edward VII and Queen Alexandra from a West Country visit. The first non-stop down trip also carried Royalty, appropriately, as the train was a relief to the *Cornishman*, the Duke and Duchess of Cornwall. With some ordinary passengers as well, it covered the 245¾ miles by the Bristol avoiding line in 3 h. 53½ m. The date was 14 July 1903; four days later the regular *Cornishman* began reaching Exeter in 3½ h., Plymouth in 4 h. 50 m. and Penzance in 7 h. 46 m.

This paved the way for the *Cornish Riviera Express*, introduced non-stop to Plymouth (water troughs were specially laid at Exminster) on 1 July next year, and bringing Penzance only 7 h. from London. (Today's time is 6 h. 35 m.) During its first two years the 'Limited', as it came to be called because the load was not allowed to exceed six coaches, ran in summer only, but in 1906 it became all-the-year-round, and of course took the Castle Cary route. Though this shortened the journey, Paddington–Plymouth (performed in 4 h. 7 m.) remained the world's longest non-stop run for another twenty years.

Considerable argument surrounded the choice of locomotive, many still favouring the great single driving wheels though these were admitted to be unsuitable on the hills west of Newton Abbot. Several classes of 4-4-0s had followed each other in quick succession between 1895 and 1900 – first the 'Devons', or the

'Dukes' as they became better known, with 5 ft 8 in. drivers specially for work west of Newton Abbot, then the 'Badmintons', the famous 'Bulldogs', the 'Camels', another class specially designed for the far West, and then the 'Atbaras' and the 'Cities'. Variety indeed, though all these classes were of the same basic family, mostly with double frames, outside cranks and outside framed bogies. Half-way through the 'Dukes' the domeless boiler appeared. To the disgust of the supporters of the singles, from the start the 'Limited' was entrusted to the 'Cities' and 'Atbaras'. Indeed, weights of trains increased so much that not only did the singles quickly disappear altogether, but the 4-4-0s were also rapidly eclipsed by the 4-6-0s, heralded by the forerunner of the 'Saints' in 1902. Some of the later 'Saints', as also the 'Stars' introduced in 1906, survived until after nationalization.

After 1906 the long-distance train service quickly settled to substantially what it has remained ever since, except during the two war periods. Corridor coaches became universal on the better trains, and by 1913, for instance, eight restaurant cars a day were running on the Exeter–Newton Abbot section in summer. Until 1903, meals had been for first-class passengers only, but now most restaurant car as well as other business was third class, second being abolished in 1910. A new feature in 1913 was the *Devon and Cornwall Special*, for Paignton, Newquay and Falmouth. Though third class only it was a modern train with restaurant car, bicycle stands and reserved seats, a far cry from the Plymouth Cheap. Reserved seats were a GWR innovation, first used on the *Cornish Riviera* in 1905.

But if the basic features of the time-table were established before the war, passenger traffic grew infinitely more rapidly after it. Numerous extra trains, particularly to Torbay, began running at summer week-ends, though not until the 1930s did the summer Saturday time-table have to be printed separately. That the West Country increased its popularity more quickly than other holiday areas was due in no small measure to the GWR itself, the only Company to retain its identity after the 1923 'grouping'. Its permanence in an unsteady age earned it the literal adoration of thousands of families. The permanence was real: even among the old companies the GWR had adhered to its policies more doggedly than any other, though there had been one surprising alteration.

In 1909 the traditional chocolate and cream coach colours were discarded for dull all-over brown, followed by crimson lake. But in 1923, just as the liveries of dozens of companies began to vanish, back came chocolate and cream to gratify the public. The same year brought the 'Castles', mechanically well ahead of anything else of that date, but in physical appearance evolved directly from the smaller 'Saints' and 'Stars' and even earlier 'Dukes' and other four-coupled engines – retaining a reminder even of broad-gauge design.

The GWR was not slow to cash in on the love it inspired. Bookstalls were filled with chatty books about the railway itself for 'Boys of all ages', and guides to 'Smiling Somerset', 'Glorious Devon' and the 'Cornish Riviera' romped through editions. Timetable features such as Paddington–Torquay in $3\frac{1}{2}$ h. by the *Torbay Express* (or *Limited* as it was then officially called), once the 'Castles' had proved their paces, were introduced largely to please those who had come to treat the journey to the West as one of the most enjoyable parts of the holiday. There was great publicity value in such through services as Penzance–Aberdeen.

Though ever since 1890 it had been a sound railway, it was on this wave of popular enthusiasm that the GWR achieved its final greatness. True, until the end a vein of meanness persisted: for example, though financially the least harassed in the 1923–47 'grouping' era, unlike the other railways it made third-class passengers continue to sit four a side even on the best trains. No arm rests were provided. But then, with the broad-gauge legacy, GWR coaches were wider than others as well as generally cleaner (before 1939 engines used only first quality Welsh anthracite) and smoother running, track maintenance being impeccable. The main-line service of the 1920s and 1930s was excellent – but the old conservatism prevented prompt enough introduction of new trains to meet road competition in individual instances. When Torquay asked for a luxury train it was given one which ran on summer Mondays and Fridays only (not Saturdays) at not very convenient hours and slightly slower than the ordinary *Torbay Express*. The eight Pullmans may have had beautiful interior decorations, but they were scarcely in keeping with the simple demands of twentieth-century holidaymakers. After running but two seasons (1929 and 1930) the *Torquay Pullman* was withdrawn, failure having been apparent from the start. Once just one passenger was carried.

But good and bad are inextricably mixed and certainly both the worst and best points of the GWR sprang from the same source. It had been said in broad-gauge days: 'It is a very solid line and makes its progress in a solid style, doing some great things and many small, but all alike with the immovability of Jove.' This remained at least partially true even in the halcyon period.

CONCLUSION

What was true of the GWR largely remains true of the Western Region at the time of writing.[1] In 1958 the *Cornish Riviera Express* is, as a generation ago, composed of chocolate-and-cream coaches hauled as far as Plymouth by a 'King'. Direct descendants from the 'Castles' in the long family of GWR engines, the 'Kings' were introduced in 1927 and have had an incredibly long heyday. Though the coaches are now of British Railways standard design and do not take advantage of the wider tunnel clearances, the third- (today termed second) class compartments still lack the arm rests found elsewhere, and still seat four a side. Though upper quadrant signals were introduced on all other lines before the war, on the Western signals remain almost exclusively lower quadrant. Moreover, much of the old spirit among the staff has survived – which has proved a mixed blessing in the age of standardization. But at least British Railways agreed to rescind the decision to abolish the automatic train control extended to all GWR main lines between the wars. By means of a ramp on the track and an electric circuit, an audible indication of the distant signal is given to the driver and if necessary the brake is partly applied. Now a somewhat similar system is being installed on trunk lines of other regions, though eventually the new electronic device not involving actual physical contact will replace the original GWR installation on Western routes.

The GWR dominated the West, and the character of the LSWR with its brown-and-salmon livery and of the green Southern must be left mainly for Vol. II. By the turn of the century the LSWR

[1] This section has deliberately not been revised for later editions. See Postscript.

had become a good average railway, missing the GWR's heights but also avoiding the abysmal depths found on some of its branch lines. This was so even though the contrast between the east and the west of Exeter was marked. Waterloo to Exeter was soon run in little more than 3 h. (today's time is 2 h. 58 m.), the switchback gradients witnessing some fine locomotive work. But beyond, on 'the withered arm', only the Plymouth portion of trains ran at anything approaching express speeds, journeys to Torrington or Padstow seeming nigh interminable.

Here geography has always been the great difficulty. While the Western main line goes through South Devon and Cornwall like a fish's backbone, fed by short branches on either side, and thus has sufficient traffic for numerous expresses, the Southern splits into a series of semi-main lines, none warranting fast or frequent trains. The *Atlantic Coast Express*, for instance, conveys through carriages for Ilfracombe, Torrington, Bude, Padstow, and Plymouth as well as those for Sidmouth and Exmouth detached at Sidmouth Junction. Except on summer Saturdays the restaurant car goes no farther than Exeter, though the *Cornish Riviera*'s, of course, is busy all the way to Penzance.

The Southern, too, introduced a Pullman service – the *Devon Belle*, started in 1947, which at first ran to both Plymouth and Ilfracombe. Though widely advertised and including an observation saloon, it was thinly patronized and before disappearing at the end of the 1954 season it had been reduced to an Ilfracombe-only train running just at summer week-ends. Its unfortunate career included an accident – damage was superficial – caused by the streamlined casing of its locomotive catching that of another when passing at speed. Streamlining appeared at the end of the Southern's career, first with the 'Merchant Navy' and then the smaller 'West Country' classes. The love of experiment – GWR fanatics called it a lust for change for change's sake – remained in the blood. That the casings are now being removed is a reminder of the modifications so frequent in LSWR days.

It is strange how two different railways sharing much of the same territory remained so true to their different characters. Dieselization, of course, will end much of the individuality. That, and not nationalization, will eventually terminate the era. I say eventually, but as I write some main-line diesel locomotives are beginning work in the South West, the first area earmarked for the complete elimination of steam. In 1938, incidentally, the GWR toyed with a scheme for electrification west of Taunton.

The end is bound to be sad from various points of view. For many people the romance of steam lies in the relationship of man and machine: the way in which 'Kings' and 'Castles' and 'West Countrys' battle up gradients almost reveals the driver's character. With diesels, on the other hand, maximum power is obtainable with indifferent skill at the mere turn of a handle. But it goes deeper than that. Steam has faithfully served the West Country for nearly 130 years. Through it the railways developed new industries, new ways of life. Such continuity of work in itself cannot fail to evoke nostalgia. Yet from the economic point of view, the disappearance of steam locomotives, like the closure of branch lines, must be taken as much for granted as the circumstances which originally led to their building.

That, of course, is not to say that many branch lines would not be more prosperous today had a saner and more energetic policy been pursued in recent years. Where raising income was impossible at least costs could have been reduced. Why, for instance, should a quiet terminus such as Moretonhampstead have required a staff of seven? The signal box, as those at many other stations, could have been abolished years ago. But this is controversial ground, and despite anger at the utter lack of interest in branch lines shown by the higher officials – scarcely ever do they even trouble to make a last trip themselves before the tracks are removed – it must be remembered that capital expenditure to increase efficiency and particularly to reduce labour costs was more or less prohibited for about fifteen years. It is obviously right that the chief modernization programme should be directed to the main lines, though even an Act of Parliament enabling branch-line level-crossing gates to be removed (as they are in France) some years ago would have made a worthwhile difference, as indeed would certain minor timetable revisions. Too many connections missed by a few minutes, too many awkward departure times for commuters (there was no train from Newton Abbot to Moretonhampstead between 4.35 and 6.5 p.m., for instance) have persisted year after year. The rise and fall of rural public transport must be the subject of another work.

At the end of my present story I am fully conscious of shortcomings in portraying the age. Space itself has been the enemy, in these days of high printing costs. I should have liked to describe some of the early signals, disc and crossbar or fantail affairs, wholly unlike the simple semaphores of later days; the careers of the principal freight trains, from the broad-gauge Paddington–Plymouth *Tip*, which with 25 loaded waggons averaged over 30 m.p.h. and was the world's fastest goods, down to its vacuum-brake-fitted modern counterpart travelling at express speeds; the running of the Paddington–Penzance travelling post-office, one of the country's two exclusively mail trains; and the work of the operating department on those hectic summer Saturdays when goods and locals make way for the streams of holiday expresses. But while many other people have written on the operating angle, my chief concern has been to depict the railways against the background of the country they served, and if I have stimulated a new appreciation for even a few readers the labour will have been rewarded.

CHAPTER XII

Postscript

In historic terms railways are still a modern innovation. Thus the years between the completion of the first manuscript of this book in 1958 and the revision for the fourth edition in 1973 account for well over a tenth of the total time since trains first steamed into the West. Moreover, it has been a fifteen years of dramatic upheaval. At the risk of labouring the mechanics I have therefore decided to retain the postscripts to the 1963 and 1966 editions as statements in their own historic right and to add a further note for this edition.

1963

When I wrote this book, in the first half of 1958, 'King' class locomotives were still hauling a chocolate-and-cream *Cornish Riviera Express* on a 4-hour non-stop run from Paddington to Plymouth, 0-6-0 goods engines were shunting at country stations at precisely the same times of day as they had done fifty or sixty years earlier, and in Devon and Cornwall much of the old rivalry persisted between the Western and the Southern. Neither the war nor nationalization had seemed to leave more than superficial marks; the permanence of the railways in the unsteady 1950s was both assumed and decided.

But the permanence was illusory. Steadily large parts of the system were becoming an anachronism. Losses mounted and morale sank. Many of the staff as well as the public they serve are now paying dearly for the stagnation of the first fifteen years after the war. Much of this might be said of British Railways as a whole, but in the West Country especially time stood still. Cornwall and North Devon did not lose a single branch line.

Boundary adjustments and poor relationships between Regions have been wasters of initiative. Apart from Birmingham (and London), Exeter was the only place in which two Regions had district offices, and such was the lack of liaison that Western and Southern planned their diesel multiple-unit train schemes for South and East Devon respectively in complete isolation. At one time no less than three Regions had a hand in the affairs of the Somerset & Dorset. The GWR route to Weymouth south of Castle Cary was transferred to the Southern, which promptly began erecting upper-quadrant signals, although engine drivers continued to be Western men; now the line between Castle Cary and Dorchester is back in the Western fold. Further west, the commercial affairs of the Southern's system beyond Exeter – 'the withered arm' – were for a number of years transferred to the Western, although the Southern continued to operate it; in 1958 the Southern regained complete control. The closure of the Western Region branch from Plymouth to Tavistock was then delayed because the Southern would not at first take over sections retained for goods between Lydford and Tavistock and Lydford and Lifton. No attempt was made to draw up a long-term plan for North Cornwall, which might have favoured retaining Lydford–Launceston as a passenger branch instead of the longer and more costly Halwill route. The Southern jealously retained its independent route through Devonport, spurning the cheaper alternative of handing traffic to the Western at St Budeaux. Then, on 1 January 1963, not only the withered arm but all Southern lines west of Salisbury were transferred entirely to the Western Region.

Even within the Western Region, administrative changes have led to discontent and waste. The Exeter and Plymouth districts lost the right to make their own timetables and decide many routine matters when a divisional office was set up in Bristol. A new south-west division, based on Plymouth, was established on 1 January 1963, embracing the Western's Exeter and Plymouth districts and the Southern's Exeter district. At the time of writing many administrative staff at Exeter and Plymouth face redundancy, loss of status or a move. When the dust has settled, having all railways in the area under the same local divisional management should lead to superior efficiency; but the new offices would have fitted better in Exeter than in Plymouth. Indeed, there lingers a suspicion that some of the changes have been made especially to justify the costly nine-storey administrative block opened at North Road in 1962.

Another big change, of course, has been the dieselization of the Western Region. This, too, may prove rewarding in time to come, but initial results have been disappointing, largely because of a high breakdown rate with the 'Warship' diesel hydraulic class. A quarter of the fleet has sometimes been out of commission with transmission and other faults. This, coupled with the late delivery of locomotives, has postponed the elimination of steam, and even at the beginning of 1963 expresses as well as freight trains are sometimes steam-hauled to Plymouth or Penzance.

In 1957 and 1958, when the 'Kings' and 'Castles' started showing signs of age, the public were promised big improvements with dieselization. Higher average speeds are indeed maintained on some sections but, especially on the Paddington–Plymouth run, the best trains are now slower because of additional stops. As from September 1961, when time-honoured services were 'rationalized', the *Cornish Riviera Express* has taken an extra 15 minutes to Plymouth and the *Torbay Express* an extra 25 minutes to Torquay. All trains now carry through coaches for both Torbay and Plymouth, but Plymouth passengers in the new through coaches off the *Torbay Express* arrive later than when they had to change at Exeter.

Ironically, on the very day that the diesel-hauled Western expresses were slowed down, the Southern speeded the *Atlantic Coast Expresses* by 7 minutes between Waterloo and Exeter – using steam power. The 171¾ miles are now run in 2 h. 35 m. Bulleid's 'Merchant Navy' class locomotives, rebuilt without streamlined casings, to the disgust of the designer, now retired, have served nobly. Waterloo–Exeter is Britain's last trunk route to be exclusively steam-operated; and steam's swan song has earned considerable patronage at the expense of the Western line.

Although the Western might naturally be loath further to lower the importance of Paddington (Weymouth is no longer served from there, and most Birmingham traffic will cease when the London Midland Region's electrification is completed), it has been suggested that London–Exeter daytime passenger traffic may eventually be concentrated at Waterloo. Night and freight trains would travel from Paddington *via* Bristol. The retention of three routes to the West seems unjustified, and there is little intermediate traffic between Reading and Taunton.

Whatever the future of the trunk lines, it will fall to the Western Region to thin out, if not completely to destroy, the sprawling Southern system west of Exeter. Traffic here has fallen rapidly, but so far few economies have been made. Quite apart from the general policy of reducing British Railways' mileage, the matter must crystallize when the purchase of diesel locomotives and multiple units to replace steam power becomes necessary. The high cost of dieselization has, of course, been responsible for many Western branch closures. One cannot resist saying that had diesel units been introduced on branch lines and local services in the 1930s and immediately after the war, and had signalling and other economies been vigorously pursued, the position would be much brighter today. Obvious economies and adjustments are shunned year by year on lines still open, for the reason that soon they will be closed.

Railways in the South-west are in low water. Whole chapters could be written about muddles concerning individual trains. To quote one example, the *Cornishman*, one of two through trains from Birmingham, stops at Weston-super-Mare only to pick up passengers. Passengers from Birmingham to Weston-super-Mare (places connected by substantial industrial and other ties) have to change at Bristol. Much main-line as well as local traffic has been lost even since the rest of this book was written. Nightly milk traffic from Devon to London has been transferred to the roads at the very time that British Railways have been talking about winning back long-distance freight. West Country 'exiles' increasingly come home for weekends by car, and four adult holidaymakers find hiring a car cheaper than paying 3d a mile each in fares, though they may still send some luggage by train.

Against this, mail and newspaper services are as efficient as ever – the whole South-west's business life is disorganized on the rare occasions that the down travelling post office is late – sleeping-car traffic is increasing, and newly retired residents moving in from other parts of the country choose homes within easy reach of stations lest they should need hastily to return to civilization. The West still depends heavily upon its railways, as was demonstrated during the severe winter of 1962–63. And given a new spirit of enterprise – of which we can possibly see the first signs – our children and our grandchildren may train-spot at places such as Teignmouth's and Dawlish's sea wall in the twenty-first century.

They will not, however, see such interesting trains. In the last four and a half years we have lost Western's chocolate-and-cream livery for named expresses, the ocean specials from Plymouth, the non-stop Plymouth–Paddington runs, the *Pines Express* on the Somerset & Dorset, the last push-and-pull auto-car trains, and many locomotive classes, including the veteran Beattie and Adams tanks which gave such a nice touch to the railway scene at Wadebridge and Lyme Regis respectively. There is one small newcomer: as I write it is announced that *Woolwich*, an 18-in.-gauge 0-4-0 contraption with outside pistons, built in 1916 and for many years employed in hauling ammunition at Woolwich arsenal, is at Easter 1963 to begin a new career taking 20-minute trips at 3 m.p.h. through the show gardens at Bicton, 2½ miles from Budleigh Salterton.

1966

The pace of change and decay has not slackened in the three years since the second edition of this book was published. The biggest change, perhaps, was the relegation of the former Southern Railway's main line in September 1964. Since then all passenger traffic between London and Devon and Cornwall, excepting that for East Devon, has been concentrated on the Western's Paddington service. Only semi-fast trains now use the Salisbury–Exeter section of the former Southern route, and none of the day-time trains from Waterloo goes further west than Exeter. On 7 March 1966 many of the intermediate stations between Salisbury and Exeter were closed, including Seaton Junction and Templecombe, for both the Seaton branch and the entire Somerset & Dorset system disappeared from the map the same day. Sidmouth Junction and the Sidmouth and Budleigh Salterton lines are expected to go later in 1966. By the end of the year, parts of the

Salisbury–Exeter route will be reduced to single track, while already the former Southern system west of Exeter – the 'withered arm' – is only a shadow of its former self.

Throughout the area covered by this book, the concentration of goods traffic on a few selected centres has continued apace. Small goods stations have been closed by the dozen, and even some larger, modern ones such as Paignton have gone. Soon no coal is to be handled at any station between Exeter and Plymouth, and though Newton Abbot will continue to deal with general freight traffic, its Hackney marshalling yard is to be removed. Much of this concentration is certainly desirable and some seriously overdue, but at present it seems unlikely that it will help stem the rapid decline in the railways' general freight.

Further closures are to come, and indeed nobody can yet safely forecast just how much of the system will eventually survive. Changes of policy at government and Regional level have continued to cause much waste of effort; the Western Region abolished the Plymouth Division in less than three years after establishing it. Those three years saw three general managers at Paddington and four chief officers (with no less than three different types of status) at Plymouth.

There are, of course, some positive achievements. A faster passenger service from Paddington is helping both the railways and the South West. Electric trains will soon reach Bournemouth, at present the last stronghold of steam power: steam disappeared from the Western Region with the closure of the Somerset & Dorset system. And special trains now carry Cornish china clay to the potteries. But given only stability, much more might have been done. Alas, even the unkempt appearance of the trains themselves seems to reflect the general lack of heart.

1973

Each Friday evening several thousand people, mainly West Country men and women, boys and girls, with jobs in London, leave Paddington by a string of six expresses after 4.0 p.m. for Somerset, Devon and Cornwall, most of them returning by a similar number of trains on Sunday afternoon and evening. Most of the trains run to faster schedules than the GWR ever provided, contain modern rolling-stock including restaurant cars, and indeed form one of the West Country's most valuable institutions – as witness the large number of people of all kinds and from every part of the South West who converge at stations to meet the 'exiles' or welcome other weekend visitors. Save that one of the Friday-only trains still offers only a dirty buffet with grudging service, it all goes very much as it should do, and as many of us were saying it should a decade and more before it happened. Among other improvements, the last up-trains on Sunday evenings run close together and are really fast, taking the Westbury and Frome bypasses because the signal boxes at their beginnings and ends are now opened for this business – after a period of false economy when all trains had to wind their way round both loops, and when also on Sundays every single train of a scant service in both directions stopped at Dawlish and Teignmouth. The railways now understand what their weekend customers want and try to provide it – and make a handsome profit out of doing so. Yet only eight years ago a senior railway official said there was little hope of ever providing a reasonable standard of journey for people returning to Paddington on Sunday evenings because 'everyone wants to use the last train', while the previous one had many unfilled seats. (In fact the official insisted on sending his own daughter on the penultimate one, and that she disliked being thus disciplined was hardly surprising since it meant leaving home a good hour and a half earlier.) The thought that there might be two or more 'last' services close upon each other because people going to the West for the weekend would naturally wish to leave their return till the last possible moment had apparently just not occurred.

At one time indeed the management was preaching that services would be provided when it was convenient and economic to run them, not to meet especial demands. Thus instead of the long-promised improvements, when the continuation of the basic GWR traditional service was at last abandoned upon completion of dieselization, a very standard, dull timetable was forced upon the West of England. Trains left Paddington at two-hourly intervals, all carrying through coaches for both Plymouth and Torbay, there being no concessions to any special traffic demands such as businessmen wishing to return home smartly after a day's work in London. And the traffic did actually dwindle at a time it could have been substantially increased.

So, alas, the muddles and miseries mentioned in the 1963 and 1966 epilogues continued to haunt the system for years more, but as witness the improvements on Friday and Sunday evenings for Londoners spending a weekend in the West of England, gradually sense began to dawn. After the period when nobody seemed to care, mistakes were still being made but at least attempts were made to rectify them. Finally it may be conceded there is a more imaginative as well as lively management overall. But while the West now has by far its best express service ever, making the best possible use of the locomotives and rolling-stock yet properly adapted to meet the public's requirements, the numerous changes made in getting to this happy position have inevitably taken their toll. In particular local traffic patterns have often been unnecessarily disturbed or killed outright. To cite one example, when Paddington to Plymouth trains ceased to run at half-past even hours and were changed to run at half-past odd hours, inevitably some local traffic patterns were upset in the far west, where many people were still using the expresses just to travel (maybe home from work) from one station to the next. But omitting a Newton Abbot stop from the 3.30 p.m. from Paddington left a totally unnecessary gap in the Newton Abbot to Plymouth service from 3.58 to 8.23 p.m. 'We are not interested in local traffic,' tended to be the reaction whenever such a gap was questioned; yet Torbay and mid-Devon have a combined population of 150,000 and carrying people to Plymouth, Truro and even Penzance along the same main line is surely more like long-distance or even inter-city than strictly local business. That fault has now been corrected as have many others; but not alas until they had taken their toll, since people who travelled from work to home had to find other means for a twelvemonth, and may not have been ready to return to the trains thereafter.

Had common sense been more highly prized during the last fifteen years, and had managers been less theory-ridden (maybe in their anxiety not to offend bosses promulgating the latest theoretic policies), the whole system in the South West would be running more profitably today. The examples cited above are from the Paddington–Penzance line, but much the same has happened elsewhere in the territory covered by this book. Undoubtedly one reason for a more stable management is the more stable physical railway, for after fifteen years of continual upheaval the map is now changing little, there indeed being few branch lines left to close and most surplus sidings and other tracks having been removed, while growth of traffic has happily ended talk of singling between Reading and Taunton and through Cornwall.

The rate of physical change was certainly a strain on management; yet changes in the pattern of management itself were probably even more telling. The 1963 epilogue welcomed the creation of the Plymouth Division; now it has disappeared, control being exercised from Bristol, and possibly later from Cardiff. In three years there were four top railwaymen in Plymouth with three different designations and levels of responsibility; and lower down the scale changes have been just as frequent, some management staff taking four or five posts in quick succession and ending up as booking-clerks. Promises made by one general manager were quickly broken by another; so though *all* railwaymen have been genuine in their desire to keep the main line through Cornwall alive, one cannot blame the public and the media for being suspicious.

Again, today's passenger express service is the best ever. It is pointless to say that the *Cornish Riviera Limited* is so many minutes quicker to Plymouth and so many more again to Penzance than its steam-hauled predecessor, since while the GWR ran only one top train on each route, today there are several – and on some routes many. On all main lines speed limits have been raised by easing out curves and improving pointwork. So throughout the area served by this book, the proportion of businessmen who go to London by train rather than car is positively rising at the time the road system itself is being improved.

But the freight picture is by no means so happy. Dr Beeching said that the future lay in liner trains. The West Country was given one, serving St Blazey and Plymouth; it failed and was withdrawn. Company and special-traffic trains carrying cars, cement, clay, fertilizers, oil and steel abound, but general goods mainly go by road. One of the follies of the decade was the opening of a too-ambitious coal concentration depot at Exeter; to make it economic all household coal for the whole of Devon and Cornwall has to be taken off the rails here and sent by lorry, some of it on journeys of more than a hundred miles. Many of the roads the lorries used are narrow, twisting and badly congested; yet the traffic was urgently needed to help keep alive railways to places such as Barnstaple. As on the passenger side, at least now the mistakes of the past have been realized, and throughout British Railways there is renewed interest in waggon-load business. Meanwhile the closing of goods stations continues, even at sizeable places like Poole, Weymouth and Torbay (Paignton). With Poole's traffic absorbed, Bournemouth's goods station is far less busy than it used to be, albeit most of the steel used in the district's modern buildings has passed through it. Goods traffic once handled at more than a dozen stations including Paignton, Teignmouth and Totnes is now concentrated at Newton Abbot, yet even though several sidings there have been taken up to form a depot for the reception of coal by road from Exeter, there is still none of the former pressure on the remaining sidings.

In the 1960s the management seemed almost to kill any interest the public had left in railways. In sharp contrast to the GWR's policy, as exemplified by its publication of books for 'boys of all ages', one Western Region general manager in fact greeted me for the first time hoping that I was not 'an enthusiast'. Another general manager removed all historic pictures from the walls at headquarters in Paddington; the railways had no history, it was decreed. Staff morale sank to a low level, not least because the railways were run as though nobody could possibly want to take an interest in any aspect of their running. Happily the worst is over. Pride in the job is steadily being restored, though not helped by further closures of administrative offices and maintenance depots, such as the Newton Abbot diesel depot which had an especially good reputation for quality. More railwaymen actively encourage traffic; an Exeter guard himself hired a train to show that cheap fares to London could make special trains pay. A commercial firm now runs almost weekly trips under contract from British Railways.

Trains look neater again in the blue-and-white livery that has now become universal for main-line stock; blue and white may be a far cry from chocolate and cream, but for a long-drawn-out object such as a train one approves the horizontal division of colour. At all events a new generation of enthusiasts gather at stations and focal points with cameras and notebooks. Of course the railways are not as interesting as formerly. One by one the more fascinating features have been removed. Mail trains no longer exchange bags with lineside apparatus at speed; it is many years since the 8.30 a.m. from Plymouth to Paddington carried a Reading slip-coach in its rear. Banking engines have virtually disappeared, even between Newton Abbot and Plymouth the load of freight trains being kept within the limit of a single locomotive. But Petrockstowe and Wenford bridge are among the more exotic places still on the railway system for mineral traffic, while enthusiasts now have their own two special lines, the Dart Valley from Totnes to Buckfastleigh and the Torbay Steam Railway from Paignton to Kingswear. Only an hour before writing this I watched several newly-painted chocolate-and-cream coaches being marshalled by a benevolent British Railways at Newton Abbot, while the familiar GW whistle again echoes around the Devon hills.

It is inevitable that when groups of enthusiasts meet they spend most time in nostalgic vein. Yet even for the boy just becoming interested in railways, the West Country continues to offer an enormous variety – from the live steam lines to fast expresses passing through glorious scenery, from a handful of picturesque branch lines still worth enjoying to many pieces of industrial archaeology such as Brunel's atmospheric towers and viaducts on long-abandoned routes.

It has to be remembered that the railways were invented and built to serve man, that a great deal of what has happened in the West of England in the past 130 years is directly the result of trains, that trains still play a vital part in the region's life, and that even where they have now disappeared their benefits often linger on. To repeat an earlier statement, there is not in the whole of the West of England a single well-balanced holiday resort that did not at one stage depend heavily on its railway station.

Appendices

APPENDIX I

The Four Broad-Gauge Companies in 1861

Statistics, taken from the Board of Trade returns, for the four main-line railways between Bristol and Penzance during a typical year of the broad-gauge era – the year ending 31 December 1861.

	Bristol & Exeter	*South Devon*	*Cornwall*	*West Cornwall*
		Length and Rolling Stock		
Length of Line open:				
Double	76	30	3	1
Single	37	47	51	41
Locomotives	60	supplied by contract	supplied by contract	11
Passenger carriages	131	84	46	29
Other passenger train vehicles	73	41	21	7
Goods waggons	910	246	102	312
Other waggons	59	18	8	nil
		Traffic		
Number of passengers:				
1st	179,240	137,054	32,858	24,116
2nd	1,029,377	448,908	162,380	66,796
3rd and Parliamentary	260,055	802,430	310,966	358,126
Season and Periodical Tickets	202	306	43	13
Carriages carried	1,508	698	221	46
Horses	4,168	1,883	477	54
Dogs	5,218	4,411	1,579	312
Cattle	21,835	5,956	3,348	1,115
Sheep	128,030	7,929	8,818	2,809
Pigs	2,090	420	1,178	182
Minerals (tons) coal	197,639 (coal and others)	29,089	7,428	66,833
others		39,075	9,718	23,981
General merchandise (tons)	303,035	151,888	39,532	34,249
Number of passenger trains	52,462	15,330	3,474	3,358
Number of goods trains	4,736	3,998	947	3,892
Passenger train miles	826,667	412,261	182,897	102,456
Goods train miles	235,664	120,209	41,823	47,825

	Bristol & Exeter	*South Devon*	*Cornwall*	*West Cornwall*
Average fares per mile (in pence):				
1st	2·28	2·57	2·64	2·33
2nd	1·70	1·72	1·77	1·67
3rd and Parliamentary	0·99	0·98	0·99	0·98
Express fares:				
1st	2·63	2·85	2·94	—
2nd	1·99	1·88	1·94	—
Gross passenger traffic receipts (including mails, etc.)	£222,587	£140,222	£53,646	£21,033
Gross goods traffic receipts	£122,709	£44,479	£14,149	£19,498

Working Expenses

	*	†		
Maintenance of way	£37,633	£23,469	£8,568	£4,643
Locomotive power	£51,293	£38,669	£15,666	£6,111
Repairs and renewals, carriages and wagons	£14,585	£9,255	£1,380	£2,279
Traffic charges	£40,004	£27,919	£8,468	£8,468
Rates and taxes	£9,356	£2,416	£724	£897
Government duty	£7,148	£3,362	£1,241	£323
Compensation for personal injury	£3	Nil	£200	Nil
Compensation for goods	£237	£692	£191	Nil
Legal and Parliamentary	£7,184	£1,981	£398	Nil

Balance Sheet

Gross receipts	£345,296	£184,701	£67,795	£40,531
Total expenses (including miscellaneous)	£181,745	£107,763	£39,290	£25,187
Net receipts	£163,551	£76,938	£28,505	£15,344
Percentage of expenses to receipts	53	58	58	62

* Including Exeter & Crediton and Somerset Central till August 31st.
† All South Devon figures including South Devon & Tavistock and Dartmouth & Torbay.

	Bristol & Exeter	*South Devon*	*Cornwall*	*West Cornwall*
		Capital		
Authorized shares	£4,000,000	£1,839,550	£1,375,060	£500,000
Loans	Nil	£605,666	£458,000	£165,000
Paid up at 31.12.61 (and dividend paid during year)				
Ordinary Capital	£2,022,460 (4⅝%)	£1,469,425 (1⅜%)	£570,459 (Nil)	£353,825 (1½%)
Preference	£1,045,372 (4%)	£100,000 (4½%) £247,290 (5%)	£400,000 (5%) £50,260 (4½%)	£6,895 (5%)
Debentures		£45,395 (4½%) £80,875 (4%)		
Loans outstanding	£214,945 (4½%) £7,300 (4¼%) £656,881 (4%) £14,650 (3½%)	£73,290 (5%) £249,405 (4½%) £10,550 (4¼%) £104,665 (4%) £500 (3½%)	£248,862 (5%) £78,441 (4½%) £25,790 (4¼%) £21,700 (4%)	£121,625 (5%) £43,330 (4½%)
Subscriptions to other companies	£449,966	£75,000	Nil	Nil

APPENDIX II

Plymouth Area Passenger Stations and Halts

This list, in conjunction with the maps for Chapter VI (and caption on page 118), illustrates both suburban and rural trends which are fairly typical. The lateness of the opening of many of the halts will be noticed. Where the date is given only to the nearest month the source is usually *Bradshaw*.

The abbreviations are: H. Halt (usually unstaffed); P. Platform (usually with staff); R. Road; S. Southern; W. Western.

WESTERN MAIN LINE

PLYMPTON (*opened* 15 June 1848, *closed from* 3 March 1959).

LAIRA H. (*opened* 1 June 1904, *closed from* 7 July 1930).

LAIRA (*opened* 5 May 1848, *closed from* 4 April 1849).

Temporary two-shed station, awaiting completion to Millbay. Station 'superintendent' paid £100 a year. There was a ban on 'spiritualistic liquor' in refreshment room. Fare on special buses into Plymouth was 6d single: often the accommodation was inadequate.

LIPSON VALE H. (*opened* 1 June 1904, *closed from* 4 May 1942).

Used by S. only after withdrawal of W. suburban trains on 7 July 1930. Closed and dismantled because of fire risk—wooden platforms.

MUTLEY (*opened* 1 August 1871, *closed from* 2 March 1939).

Down platform opened several weeks later. For some years L.S.W.R. had its own staff. Known as 'the gentry's station'.

NORTH ROAD (*opened* 28 March 1877).

Was planned as stone station but built of wood because of G.W.R.'s delay in starting it and L.S.W.R.'s impatience for its opening. The main up and down tracks had platforms on either side. Refreshment room opened 1888. Station enlarged 1908. Rebuilding started in 1938; discontinued in war; restarted in 1956 and completed with tall office block and colour-light signalling in 1962. Named just 'Plymouth' after closure of Friary on 15 September 1958.

MILLBAY (*opened* 4 April 1849, *closed from* 23 April 1941 to passengers, 20 June 1966 to goods).

Original station with 'hydraulic machinery for crossing the engine from one line of rails to the other', extended for Cornwall and Tavistock lines' traffic in 1859, and replaced by existing station in 1900. Ticket platform built outside August 1851 and closed in 1896. Shortness of Millbay's platforms had long presented difficulty and before the war most Cornish expresses by-passed it. Bombing of adjoining goods station, and need to use passenger platforms for goods, responsible for closure. But many trains starting at Plymouth continued to be formed in the station and carriage sidings between Millbay and North Road were extensively used until their *closure on* 20 June 1966. With the withdrawal of all rail traffic at Millbay Docks, nothing remains of the Millbay complex south of the main line. The direct access from Millbay to Cornwall had been abandoned a decade earlier.

WINGFIELD VILLAS H. (*opened* 1 June 1904, *closed from* June 1921).

DEVONPORT (ALBERT R.) (*opened* 4 May 1859).

Formerly just Devonport.

DOCKYARD H. (*opened* 1 June 1905).

FORD H. (*opened* 1 June 1904, *closed from* 6 November 1941).
Formerly Ford P. Not reopened after bombing.

KEYHAM (*opened* 1 July 1900). Mini-marshalling yard still in use.

ST BUDEAUX (FERRY R.) (*opened* 1 June 1904).
Formerly St Budeaux P.

SALTASH (*opened* 4 May 1859).
Rebuilt 1880 and extended after Royal Albert Bridge approach had been altered 1908. Until the opening of the Saltash road bridge, Saltash was the busiest passenger station in Cornwall with over 30 daily trains to Plymouth, some carrying several hundred originating passengers each. Much reduced, the Plymouth–Saltash suburban extras finally ceased in May 1972 though a few trains to and from points further west in Cornwall serve Saltash and intermediate stations to Plymouth. Saltash *closed* to goods on 9 September 1963.

DEFIANCE P. (*opened* 1 March 1905, *closed from* 27 October 1930).
Named after training ship H.M.S. *Defiance,* moored in river. At first called Defiance (Wearde). Rebuilt on fresh site in 1907.

SOUTHERN MAIN LINE

Section between St Budeaux and Devonport Junction *closed* completely 7 September 1964. Section from Okehampton and Lydford to Bere Alston *closed* to freight 28 February 1966 and to passengers 6 May 1968.

LYDFORD (See Launceston Branch).

BRENTOR (*opened* 2 June 1890).
Two railways run parallel from Lydford to Tavistock, and Plymouth, Devonport & South-Western Junction chose to serve Brentor instead of Mary Tavy, on Launceston branch.

TAVISTOCK NORTH (*opened* 2 June 1890).
North added after nationalization. *Closed* to freight 28 February 1966.

BERE ALSTON (*opened* 2 June 1890).
Beer Alston till 1898. Junction for Callington, and since 1968 effectively only intermediate station, though reversal point, on St Budeux–Gunnislake branch.

BERE FERRERS (*opened* 2 June 1890).
Beer Ferris till 1898.

TAMERTON FOLIOT (*opened* January 1898, *closed* 10 September 1962).

ST BUDEAUX (VICTORIA R.) (*opened* 2 June 1890).
Formerly St Budeaux (for Saltash) and just St Budeaux.

WESTON MILL H. (*opened* 1 November 1906, *closed from* September 1921).

CAMEL'S HEAD H. (*opened* 1 November 1906, *closed from* 4 May 1942).

FORD (DEVON) (*opened* 2 June 1890).
Formerly just Ford. *Closed* 7 September 1964.

ALBERT R.H. (*opened* 1 October 1906, *closed from* 13 January 1947).
In opening between tunnels.

DEVONPORT (KING'S R.) (*opened* 18 May 1876).
Formerly Devonport and Stonehouse. Terminus until new Lydford–Devonport line opened 2 June 1890, when up and down lines were reversed. There was a refreshment room. *Closed* 7 September 1964 to passengers and 4 January 1971 to goods.

(NORTH ROAD, MUTLEY, and LIPSON VALE H.—see W. Main Line.)
FRIARY (*opened* 1 July 1891, *closed from* 15 September 1958).
At its busiest in last war, with Yealmpton as well as Turnchapel and S. main line trains. Closed in order to be converted into Plymouth's main goods station under modernization plan and since closure of Tavistock Junction marshalling yard on 4 January 1973 has also been the main Plymouth marshalling yard.

LAUNCESTON BRANCH (W.)

Branch *closed* to passengers 31 December 1962. Marsh Mills, and sections between Lydford and Tavistock South and Lydford and Lifton retained for goods. Marsh Mills later turned into siding served by Tavistock Junction marshalling yard and the siding still used for clay traffic though Tavistock Junction marshalling yard closed in January 1973. Lydford–Tavistock South closed entirely 25 September 1964. Lifton–Lydford reopened for freight 7 September 1964; the same day freight withdrawn from the North Cornwall line through Launceston. Lydford–Launceston closed entirely 28 February 1966.

(MILLBAY, NORTH ROAD, MUTLEY, LIPSON VALE H.—see W. Main Line.)
MARSH MILLS (*opened* 1861).
(Marsh Mills, Bickleigh, Yelverton, Horrabridge, Tavistock, Mary Tavy and Lydford also used by L.S.W.R. 18 May 1876–1 June 1890.)
PLYM BRIDGE P. (*opened* 1 May 1906).
BICKLEIGH (*opened* 22 June 1859).
SHAUGH BRIDGE P. (*opened* 21 August 1907).
CLEARBROOK H. (*opened* 29 October 1928).
YELVERTON (*opened* 1 May 1885).
Junction for Princetown. Yelverton Siding (without passenger platforms) already in existence and Princetown trains ran on L.S.W.R. narrow gauge rail to Horrabridge. The township of Yelverton developed rapidly after 1885.
HORRABRIDGE (*opened* 22 June 1859).
WHITCHURCH DOWN P. (*opened* 1 September 1906).
Unstaffed since April 1956.
TAVISTOCK SOUTH (*opened* 22 June 1859).
South added after nationalisation.
MARY TAVY AND BLACKDOWN H. (*opened* 1 July 1865).
Mary Tavy till May 1907. Crossing station till after L.S.W.R. trains ceased to use the route in 1890.
LYDFORD (*opened* 1 July 1865).
Formerly Lidford. L.S.W.R. station built alongside, making four platforms in all. Wartime connection between two lines opened 15 November 1943.
LIDDATON H. (*opened* 4 April 1938).
CORYTON (*opened* 1 July 1865).
LIFTON (*opened* 1 July 1865).
Heavy traffic for many years until *closure* in 1966.
LAUNCESTON W. (*opened* 1 July 1865, *closed from* 30 June 1952).
Wartime connection between W. and S. opened 22 September 1943. Passenger traffic concentrated at S. station from 30 June 1952.

PRINCETOWN BRANCH (W.)

Branch *closed* 5 March 1956 and dismantled 1957.

(Yelverton)

DOUSLAND (*opened* 11 August 1883).

BURRATOR H. (*opened* 4 February 1924).

Originally Burrator P. For tourist traffic already fostered by G.W.R. buses.

INGRA TOR H. (*opened* 2 March 1936).

Famous for notice warning of snakes.

KING TOR H. (*opened* 2 April 1928).

Opened to serve group of quarrymen's cottages. There was a siding into quarry. A 1909 scheme (revived 1924) for Light Railway from near H. to Merrivale did not mature.

PRINCETOWN (*opened* 11 August 1883).

YEALMPTON BRANCH (W.)

(MILLBAY, NORTH ROAD, MUTLEY, LIPSON VALE H.—see W. Main Line.)

MOUNT GOULD AND TOTHILL H. (*opened* October 1905, *closed from* 1 February 1918).

(Plymstock)

BILLACOMBE, ELBURTON X., BRIXTON R., STEER POINT, YEALMPTON (*opened* 17 January 1898, *closed from* 7 July 1930; *reopened* 3 November 1941, *closed from* 7 October 1947).

Line reopened to accommodate Plymouth people living in country during war. Unadvertised workmen's trains began on 21 July 1941. From 3 November 1941 service was from Friary, calling also Lucas Terrace H., and not from W. main line. Closed to goods traffic 29 February 1960.

TURNCHAPEL BRANCH (S.)

Branch *closed* to passengers *from* 10 September 1951. Had also been closed temporarily because of fuel crisis January to 2 July 1951. Section beyond Plymstock closed to goods 30 September 1961, heavy cement traffic handled at Plymstock.

(Friary)

LUCAS TERRACE H. (*opened* October 1905).

Also used by W. Yealmpton trains 3 November 1941–7 October 1947.

PLYMSTOCK (*opened* 5 September 1892).

Junction for Yealmpton. See above.

ORESTON (*opened* 1 January 1897).

TURNCHAPEL (*opened* 1 January 1897).

Line for Admiralty traffic extended about ½ mile beyond passenger station.

CALLINGTON BRANCH (S.)

Section beyond Gunnislake *closed* to freight 28 February 1966 and to passengers 7 November 1966.

(Bere Alston)

CALSTOCK (*opened* 2 March 1908).

Waggon hoist to quay below used for many years. Sold 1936.

GUNNISLAKE (*opened* 2 March 1908).
CHILSWORTHY (*opened* 1 June 1909).
LATCHLEY (*opened* 2 March 1908).
SEVEN STONES H. (*opened* June 1910, *closed from* 1914).
For Phoenix Pleasure Ground excursion traffic, which was heavy. Ground closed for good at outbreak of war.
LUCKETT (*opened* 2 March 1908).
Stoke Climsland till 1909.
CALLINGTON (*opened* 2 March 1908).
Callington R. till 1909, Callington for Stoke Climsland till 1949.

PLYMOUTH LOOP AND GOODS LINES 4 FT 6IN. OR DARTMOOR GAUGE

PLYMOUTH & DARTMOOR RAILWAY (*opened* 26 September 1823).
See text for main details. Plymouth terminus was Sutton Harbour, and there was short branch to Cattewater. Dartmoor gauge taken up beyond junction with branch 1857, quickly relaid, and finally taken up 1869.
CANN QUARRY BRANCH (*opened* 20 November 1829).
At first ran from Rising Sun on above line to meet canal leading from quarry. Later canal closed and railway extended on towpath to slate quarry, busy till 1855 and worked spasmodically till 1900.
PLYMPTON BRANCH (*opened* 1833 or 1834).
From Marsh Mills on Cann Quarry branch. Carried clay traffic, and extension to pits at Lee Moor considered. Bought by South Devon Railway 1847 and *closed.*
LEE MOOR TRAMWAY (*opened* September 1854).
From near Plym Bridge on Cann Quarry branch to beyond Lee Moor village. Built—shoddily—by Tavistock & South Devon Railway as price of Lord Morley's support. *Closed* 4 October 1854 by accident at Torracombe, second of two cable-operated inclines.
Reopened 24 September 1858 after being rebuilt. Carried increasing traffic. After 1916 the tramway, together with parts of the three above lines incorporated in it to enable traffic to reach Cattewater, was sole survivor of Dartmoor gauge. Locomotives used between two inclines, horses elsewhere. *Closed* 1939; part *reopened* for two years after war; by 1958 was much overgrown, signals collapsing, rolling stock rotting on track. Clay is mainly piped from Lee Moor (where production is over 500,000 tons annually) to Marsh Mills treatment works, but it was thought that London area demand for gravel, a by-product, might lead to tramways reopening, budgeted for in Plymouth Development Plan. Horse-drawn trucks occasionally proved right to cross certain roads within City until 1960, but track finally lifted in 1961–2. Two of the locomotives are preserved.

EASTERN DISTRICT

PLYMOUTH BREAKWATER RAILWAY (*opened* August 1812).
John Rennie, engineer in charge of construction of badly needed breakwater, had difficulty in transporting large stones for foundation,

so laid 3 ft 6 in. tramway from Oreston quarry to pier. Trucks could at most states of tide run straight on to one of ten boats built for work. Eventually rails also laid on breakwater itself.

SUTTON HARBOUR BRANCH (*opened* May 1853).

From Laira. Broad-gauge rail added to Dartmoor gauge line. Rebuilt for locomotives 1856–7, but owing to Dartmoor gauge being added to Sutton Harbour again (see above) locomotives not introduced till 19 April 1869. After that the branch exclusively the South Devon Railway's; i.e. the broad and Dartmoor gauge now ran parallel on separate tracks from Laira as far as the junction of the latter's Cattewater branch. *Closed* into Sutton Harbour about 1969, the branch being left in use only to level crossing short of harbour. The first section from Laira to Friary Junction carries main freight traffic in and out of Plymouth since Friary marshalling yard became focal point on closure of main-line Tavistock Junction marshalling yard in January 1973.

FRIARY BRANCH (*opened* 1 February 1878).

Single narrow gauge line from Friary Junction on Sutton Harbour branch (gauge of which was 'mixed' as far as this) to Friary goods station.

NORTH QUAY BRANCH (L.S.W.R.) (*opened* 22 October 1879).

Leaving Friary branch just before its termination. *Closed* as unnecessary duplicate after nationalization; rails removed 1957.

NORTH QUAY BRANCH (G.W.R.) (*opened* 6 November 1879).

From Sutton Harbour branch. North Quay is at north of Sutton Pool, which had been improved.

CATTEWATER BRANCH (*opened* 1879–80).

Like Turnchapel branch, built by Plymouth & Dartmoor Railway acting as L.S.W.R. spear-head. Leaves Friary branch about ½ mile from termination; reversal necessary. Exact date unknown. Extended to present terminus in 1888. Worked by L.S.W.R. Carries heavy traffic and serves a petroleum plant and Prince Rock power stations.

LIPSON JUNCTION (on W. main line) to MOUNT GOULD JUNCTION (on Sutton Harbour branch) (*opened* 1 April 1891).

Double line to enable L.S.W.R. trains to run from North Road to Friary passenger station, opened 1 July. Sutton Harbour and Friary branches between Mount Gould Junction and Friary rebuilt as double passenger route. Still used among other things for empty passenger stock movements to sidings in Laira yard beside Embankment Road that were designed under the modernization programme for storing coaches used by former Southern route into Plymouth but taken over for most Plymouth coach storage purposes after closure of Millbay system.

LAIRA BRIDGE (*opened* 5 September 1892).

Trains crossing bridge for Plymstock travelled from Friary by Cattewater branch as far as Cattewater Junction. Bridge had been built in about 1887.

MOUNT GOULD JUNCTION–CATTEWATER JUNCTION (*opened* 17 January 1898).

A spur to enable G.W.R. Yealmpton branch trains to run to and from main W. line without touching Friary; *closed from* 15 September 1958.

WARTIME SPUR (*opened* 1914).

Link between Sutton Harbour and Cattedown branches independent of main line. Taken out March 1923.

CENTRAL AND WESTERN DISTRICTS

EDDYSTONE LIGHTHOUSE TRAMWAY (*opened* 1756).

John Smeaton, instructed to build stone lighthouse as quickly as possible to replace wooden one destroyed by fire in 1755, found weather difficult: so each course of stones first erected on mainland to ensure perfect fitting, and 'a timber road, commonly called at the Collieries, where they are used, a Rail Road', laid to assist in cartage. Four-wheel trucks ran on tracks connected by a turntable. Part of this lighthouse now stands on Hoe.

G.W. DOCKS (*opened* 1850). Later known as MILLBAY DOCKS.

Dock Co. built a short connection with South Devon Railway.

1878–81. Extensions to West Wharf and Graving Dock.

18 June 1878. Narrow-gauge access, compelled by 1874 Act under which broad gauge companies bought out Dock Co. Ocean specials began at dockside and ran non-stop out of Plymouth, the junction with the Millbay passenger station line and the dock line being placed to the north of the former. The last passenger terminal opened as recently as 1952. See Chapter XI for more details of ocean specials.

CORNWALL LOOP (*opened* 17 May 1876).

Connecting North Road with Devonport Junction, bypassing Millbay. Originally built for L.S.W.R.

STONEHOUSE POOL BRANCH (*opened* 1877).

Under Stonehouse Pool Improvement Act. Part of branch had been opened previous year. *Closed* 1966.

Passenger ocean terminal. *Opened* 9 April 1904, *closed* 1911.

KEYHAM–DOCKYARD (*opened* 1867).

Narrow-gauge access provided soon after arrival of L.S.W.R. in Plymouth in 1876.

DOCKYARD RAILWAY (*opened* 1867).

Tramway may have existed earlier. Two miles long, including tunnel linking two parts of Yard separated by Torpoint Ferry approach road. Passenger services started 1900. Half-hour service during Yard hours served six stations from Extension in north to South Yard until May 1966. Five classes of accommodation were provided in old-fashioned coaches.

SHOW SIDINGS (*opened* 1865 and 1872).

Temporary sidings to Pennycomequick for Bath and West Shows.

BULLPOINT GOVERNMENT SIDINGS (*opened* 2 June 1916 and still *in situ*).

SPUR AT ST BUDEAUX (*opened* 21 March 1941).

Link between W. and S. used on numerous emergency working occasions. Rebuilt 1964. From 7 September 1964 all trains from the former Southern Railway's Tavistock route transferred to former Western metals for remainder of journey into Plymouth. Now part of the Plymouth–Bere Alston–Gunnislake branch.

APPENDIX III

Two 1844 Journeys

(I) Daniel Gooch's report to the Great Western Railway directors on the first train – 'a special trip of Directors and others' – to run to Exeter, on 1 May 1844:

Orion Engine: Cylinders, 15 in. dia. Stroke, 18 in. Wheels, 7 ft dia. Blast pipe, $4\frac{1}{8}$ in. Weight of Engine and Tender: 34 tons. Carriages with 40 people, 21 tons. Miles run by the engine: 450. Coke consumed: 85 cwt., or 21.1 lb per mile. The coke was selected for the purpose and was of very good quality. The engine was fitted with a variable expansion gear and the steam was pumped to 70 lb. in the inch. All the bearings were fitted with soft metal.

We started as above from Paddington at 7.31 a.m., and from the engine being a little too full of water a little time was lost over the first four miles of the distance. Time was made good at the 14th milepost, and we maintained it very irregularly to Reading – easing the speed through Slough, Maidenhead and Twyford stations. The engine got plenty of steam and the time allowed was ample. I think 4 to 5 minutes might have been saved. We arrived at Reading punctually to time, and started after taking in water and oiling the engine in 4 minutes.

From here to Swindon the work was done with very great ease by the engine: 10 minutes might have been saved if it was required; that is, I think the distance to Swindon calling at Reading for water, could be run with 3 or 4 carriages in 1 h. 30 m. with ease. At Swindon the Directors and party had breakfast, and we started again at 9.31 a.m., having very easy work down to Bristol. From Bath to Bristol we were obliged to be cautious, as the 6.0 a.m. train was ahead of us only a short distance. This ground could be run safely in 48 minutes; or from London to Bristol, stopping at Reading and Swindon for water or at Steventon alone, the trip *could* be performed in 2 h. 21 m.

At Bristol we took in coke and water and took on another carriage with Bristol Directors – making the load about 30 tons. With this we started at 10.44 a.m. and ran to Taunton arriving there at 11.38 a.m. – i.e., in 54 minutes. This part of the ground was not run near as hard as it might have been, for the engine did not get steam very well and we had a danger signal given in the road which delayed us a little. We started from Taunton at 11.42 a.m., and went over the ground quietly until we arrived at Tiverton Road station at 12.6 p.m., being 179 miles. Here we overtook a very long train of passengers from Bristol and had to wait for this train in, into Exeter. Away there at 12.45 p.m. Had the road been clear we would have arrived at 12.30 p.m. with great ease; the ground from Bristol if necessary could be run in 1 h. 30 m., and allowing 8 minutes in Bristol the whole distance from London could be done in 3h. 59 m. The engine arrived at Exeter perfectly cool and free from dust.

On account of the turntable not being ready the engine had to run to Taunton and back tender first to get turned. This trip did more harm to the engine than the whole of the long run, as she got covered with dust and there was much danger of her heating in her return trip. She ought to have had her fire pulled out, had she been able to stop at Exeter all day – for it had by this time got very dirty, and little pains had been taken by the men who cleaned it.

We started our return trip with 4 carriages, leaving Exeter at 5.19 p.m., and proceeded quietly as far as Wellington. We passed Taunton 6.0 p.m., and arrived Bridgwater at 6.19, where we oiled and took in water, but we got steam very badly. We started from Bridgwater at 6.24 p.m., and had a very fast run up to Bristol, doing the 33½ miles in 36 minutes. Had we had steam it might have been done sooner, as it was frequently as low as 45 lb. in the inch and seldom 55; we, however, arrived in Bristol at 7.1 p.m., where we left a carriage and took in coke and water, starting again at 7.13 p.m. I have had the fire cleaned a little, but it was badly done, in that my steam was quite as bad up to Swindon, where we arrived with only 30 lb. at 8.7 p.m.

I got Appleby to go under the engine and see that the fire was properly cleaned, where he picked several buckets of dust and clinkers out from the sides of the box, and we started again at 8.19 p.m. We now got steam very well and ran very hard until we arrived near Wallingford Road, when we got the green signals for the train that had passed ahead and which we were to pass at Reading. This slackened our speed all the way to Pangbourne where we got red signals and had to pull up nearly to a stand. We arrived at Reading close behind the train at 9.13 p.m., and took in water, etc., starting at 9 h. 20 m. From Reading to Maidenhead Engine House was the quickest run we had – the distance 13½ miles being done in 14 minutes, including getting into speed at Reading, shutting the steam off at Twyford, and through Maidenhead station. I never ran it hard; as the engine appeared to be going through the Chalk Cutting I should say little short of 70 miles per hour, it kept a good speed into London, although a little time was lost at Slough by them having on the green lights. We arrived in London in 43 m. from Reading.

During the whole of the trip the engine ran best cutting off at 12 to 13 miles. The blast pipe should have been a little smaller—I should say ⅛ of an inch. As it was, the whole time occupied in *running* the 388 miles to Exeter and back, exclusive of stoppages and the delay in going into Exeter, was 8 h. 29 m., or 45.7 miles per hour – the down trip being 4 h. 21 m., or 44.5 miles per hour, and the up trip 4 h. 8 m., or 46.9 miles per hour. The engine arrived at Paddington quite cool and in good order.

DANL. GOOCH.

(II) Record of a rail and road journey from London to St Austell on 30 November 1844, from the diary of William Pease, who was steward to J. T. Treffry, chairman of the Cornwall Railway. The journey was in the nature of a race – the special train consisted of only one coach – as in conformity with Standing Orders this was the last day plans and sections of the proposed Cornwall Railway could be deposited with the Clerk of the Peace to ensure the Bill being introduced in the 1845 Parliamentary session:

Miles	London	left	4.50 a.m.	by the 'Achilles' engine, special train
35¾	Reading		5.50	1 hr. 0 mins. going 35½ miles
		left	5.58	stopped to take in water
77	Swindon		6.38	40 mins. going 41¼ miles
		left	6.48	took in coals
106¾	Bath		7.20	took 32 going 29¾ miles
118¼	Bristol		7.40	took 20 going 11½ miles
		left	7.53	took in coals and water

136¾	Weston-s-Mare	8.15	took 23 going 18½ miles
151¼	Bridgwater	8.32½	took 17½ going 14½ miles
163	Taunton	8.47	took 14½ going 11¾ miles
170	Wellington	8.58	took 11 going 7 miles
	left	9. 4	took in water
193¾	Exeter Station	9.25	took 26 going 23¾ miles

BY POST CHAISE

	Left Exeter		10.50	
12	Croggan Well	11.59	12. 7 p.m.	12 miles in 1 hour 9 mins.
11½	Oakhampton	1. 2	1.13	11½ miles in 55 mins.
20	Launceston	3. 0	3.13	20 miles in 1 hour 47 mins.
22	Bodmin	5. 3	5.13	22 miles in 1 hour 50 mins.
12	St. Austell	6.20		12 miles in 1 hour 7 mins.

Cost of posting from Exeter to St. Austell, 4 horses & 2 postillions.

Exeter to Croggan Well. 12½ miles. £1. 17. 0. Gates 5s. Post boys 8s. £2. 10. 0.

Croggan Well to Oakhampton. 11 miles. £1. 13. 0. Gates 2s. Post boys 9s. £2. 4. 0.

Oakhampton to Launceston. 20 miles. £3. 0. 0. Gates 4s. Post boys 12s. £3. 16. 0.

Launceston to Bodmin. 22 miles. £3. 6. 0. Gates 6s. Post boys 14s. 6d. £4. 6. 6.

Bodmin to St. Austell. 12 miles. £1. 16. 0. Gates 4s. Post boys 8s. £2. 8. 0.

Total £15. 4. 6.

Bibliography and Acknowledgments

Material has been collected over many years, so a detailed list of sources has to be omitted. As those familiar with the subject will not need telling, the most useful source-book has been E. T. MacDermot's *History of the Great Western Railway* (1927 and 1931), still the standard work, though dealing only with the academic story. C. F. Dendy Marshall's *A History of the Southern Railway* is also a standard work, though research has shown that many dates are inaccurate. Where my dates differ from MacDermot's or Dendy Marshall's, it is as the result of irrefutable evidence, usually contemporary. All dates for the earlier editions were checked by Mr C. R. Clinker, whose other help was likewise invaluable. A large proportion of the material besides dates has been drawn from contemporary newspapers, timetables and official notices, directories, guides and censuses.

The list of other printed sources must be headed by the *Railway Magazine*. Unfortunately there is no master index at present. Further railway periodicals of various dates consulted are the *Railway Times*, *Railway Gazette*, *Railway World*, *Trains Illustrated* and the magazines of the Great Western and Southern Railways and the corresponding Regions of British Railways. Other books on transport subjects I used for the first edition were: *A History of the Great Western Railway* by G. A. Sekon, 1895; *The History of a Railway* (Salisbury and Yeovil), by Louis H. Ruegg, 1878; *The Somerset and Dorset Railway*, by D. S. Barrie and C. R. Clinker, Oakwood Press, 1948; *The Lynton and Barnstaple Railway*, by L. T. Catchpole, Oakwood Press, 1949; *The South Western Railway* (mechanical history), by Hamilton Ellis, Allen and Unwin, 1956; *Four Main Lines*, by Hamilton Ellis, Allen and Unwin, 1950; *Nineteenth Century Railway Carriages*, by Hamilton Ellis, Modern Transport, 1949; *Isambard Kingdom Brunel*, by L. T. C. Rolt, Longmans, 1957;

Red For Danger (accidents and safety precautions), by L. T. C. Rolt, Bodley Head, 1955; *The Great Western Railway* (an appreciation), by O. S. Nock, Heffer, 1951; *From Trackway to Turnpike* (an illustration from East Devon), by Gilbert Sheldon, Oxford, 1928; *West Country Passenger Steamers* (Bristol Channel), by Graham E. Farr, Tilling, 1956; *Somerset Harbours*, by Graham Farr, Christopher Johnson, 1954; *Dorset Harbours*, by Donald Payne, Christopher Johnson, 1953; *Devon Harbours*, by Vernon C. Boyle and Donald Payne, Christopher Johnson, 1952; *Cornwall Harbours* by Roland Roddis, Christopher Johnson, 1951; and *The Canals of Southern England*, by Charles Hadfield, Phoenix House, 1955.

Some extra material for the later editions has come from the new generation of railway histories, many of them published from Newton Abbot station. They include *The London & South Western Railway*, Vol I, and I have also used the manuscript of Vol II, by R. A. Williams; *The Somerset & Dorset Railway*, by Robin Atthill (also by the same author *The Picture History of the Somerset & Dorset Railway* and a more specialist work by a number of authors on *The Locomotive History of the Somerset & Dorset Railway*); *The Midland & South Western Junction Railway*, by Colin G. Maggs; *The Haytor Granite Tramway and Stover Canal*, by M. C. Ewans; *The Lynton & Barnstaple Railway*, by Brown, Prideaux and Radcliffe; *Atmospheric Railways*, by Charles Hadfield; and among works on the locomotive side, *The GWR Stars, Castles and Kings* (two volumes) and *GWR Steam*, all by O. S. Nock. *Summer Saturdays in the West* by myself and Simon Roxborough Smith deals in detail with the handling of peak passenger traffic.

The West Country volume in the 'Railway History in Pictures' series by P. Riley complements this work in a pictorial sense, and Charles Hadfield's *The Canals of South West England* portrays the canal age in much the same territory. Many works have been published on industrial history and archaeology since this work was first published. Examples are *The Industrial Archaeology of the Tamar Valley*, by Frank Booker, and uniform volumes on Cornwall by A. C. Todd and Peter Laws and on Dartmoor by Helen Harris who has also written individual histories of the Bude and Grand Western canals. Canal titles by

Kenneth R. Clew of special interest to railway historians are *The Somersetshire Coal Canal and Railways* and *The Kennet & Avon Canal*. But one could expand this list endlessly and my own firm of David & Charles at Newton Abbot and Bradford Barton at Truro each publish new titles on the West Country's transport and industrial history every year. When one reflects that when this present book was first published in 1960 it was only the third or fourth book on any aspect of industry, commerce or transport in the West of England to appear after the war, the increasing interest in these topics is obvious. In fact in 1960 no other book contained half so much material on West Country industry as did this. The growth in newspaper, magazine and journal articles of merit has been equally marked.

In preparing the original edition I consulted a wide range of general books on individual towns, districts and counties, but in view of later titles that have again broadened the literature available, there is little point in mentioning individual works. However, the work of one author should not be omitted. A. K. Hamilton Jenkin is the doyen of Cornish historians, and his *The Cornish Miner* and *Cornwall and Its People* vividly demonstrate the big role played by the county in the early days of the industrial revolution. Many maps were also consulted including the first edition of the one-inch ordnance survey, now reprinted.

Literally hundreds of people have contributed material and other help for the various editions of this book, ranging from librarians, one of whom I married, to railwaymen of all ranks. The following list is a select one of people who actually placed the fruits of their own research at my disposal: Messrs M. D. Greville, Ivor Hocking, John Keast, H. G. Kendall, Ashley Rowe, R. C. Sambourne, Roger Selleck and R. J. Woodfin. Finally that librarian I married has not only provided much encouragement and done the typing and checking now on four occasions, but has kept an eye on the work on behalf of readers who may be new to railway history.

Index

Bold type is used to indicate pages giving opening dates of sections of line (not individual stations). Main line opening dates are indexed under the company's name; branch openings are indexed under the station at the furthest end, or in some cases under the popular name of the branch. All references to suburban lines and stations are indexed under the town or city concerned. Principal industries and traffics are indexed individually.